T0212281

Analysis and Design of Transmitarray Antennas

Synthesis Lectures on Antennas

Editor

Constantine A. Balanis, *Arizona State University*

Synthesis Lectures on Antennas will publish 50- to 100-page publications on topics that include both classic and advanced antenna configurations. Each lecture covers, for that topic, the fundamental principles in a unified manner, develops underlying concepts needed for sequential material, and progresses to the more advanced designs. State-of-the-art advances made in antennas are also included. Computer software, when appropriate and available, is included for computation, visualization and design. The authors selected to write the lectures are leading experts on the subject who have extensive background in the theory, design and measurement of antenna characteristics. The series is designed to meet the demands of 21st century technology and its advancements on antenna analysis, design and measurements for engineers, scientists, technologists and engineering managers in the fields of wireless communication, radiation, propagation, communication, navigation, radar, RF systems, remote sensing, and radio astronomy who require a better understanding of the underlying concepts, designs, advancements and applications of antennas.

Analysis and Design of Transmitarray Antennas
Ahmed H. Abdelrahman, Fan Yang, Atef Z. Elsherbeni, and Payam Nayeri
2017

Design of Reconfigurable Antennas Using Graph Models
Joseph Costantine, Youssef Tawk, and Christos G. Christodoulou
2013

Meta-Smith Charts and Their Potential Applications
Danai Torrungrueng
2010

Generalized Transmission Line Method to Study the Far-zone Radiation of Antennas under a Multilayer Structure
Xuan Hui Wu, Ahmed A. Kishk, and Allen W. Glisson
2008

Narrowband Direction of Arrival Estimation for Antenna Arrays
Jeffrey Foutz, Andreas Spanias, and Mahesh K. Banavar
2008

Multiantenna Systems for MIMO Communications
Franco De Flaviis, Lluis Jofre, Jordi Romeu, and Alfred Grau
2008

Analysis and Design of Transmitarray Antennas

Ahmed H. Abdelrahman, Fan Yang, Atef Z. Elsherbeni, and Payam Nayeri

ISBN: 978-3-031-00413-1   paperback
ISBN: 978-3-031-01541-0   ebook

DOI 10.1007/978-3-031-01541-0

A Publication in the Springer series
SYNTHESIS LECTURES ON ANTENNAS

Lecture #12
Series Editor: Constantine A. Balanis, *Arizona State University*
Series ISSN
Print 1932-6076   Electronic 1932-6084

Analysis and Design of Transmitarray Antennas

Ahmed H. Abdelrahman
University of Colorado Boulder

Fan Yang
Tsinghua University

Atef Z. Elsherbeni
Colorado School of Mines

Payam Nayeri
Colorado School of Mines

SYNTHESIS LECTURES ON ANTENNAS #12

ABSTRACT

In recent years, transmitarray antennas have attracted growing interest with many antenna researchers. Transmitarrays combines both optical and antenna array theory, leading to a low profile design with high gain, high radiation efficiency, and versatile radiation performance for many wireless communication systems. In this book, comprehensive analysis, new methodologies, and novel designs of transmitarray antennas are presented.

- Detailed analysis for the design of planar space-fed array antennas is presented. The basics of aperture field distribution and the analysis of the array elements are described. The radiation performances (directivity and gain) are discussed using array theory approach, and the impacts of element phase errors are demonstrated.

- The performance of transmitarray design using multilayer frequency selective surfaces (M-FSS) approach is carefully studied, and the transmission phase limit which are generally independent from the selection of a specific element shape is revealed. The maximum transmission phase range is determined based on the number of layers, substrate permittivity, and the separations between layers.

- In order to reduce the transmitarray design complexity and cost, three different methods have been investigated. As a result, one design is performed using quad-layer cross-slot elements with no dielectric material and another using triple-layer spiral dipole elements. Both designs were fabricated and tested at X-Band for deep space communications. Furthermore, the radiation pattern characteristics were studied under different feed polarization conditions and oblique angles of incident field from the feed.

- New design methodologies are proposed to improve the bandwidth of transmitarray antennas through the control of the transmission phase range of the elements. These design techniques are validated through the fabrication and testing of two quad-layer transmitarray antennas at Ku-band.

- A single-feed quad-beam transmitarray antenna with 50 degrees elevation separation between the beams is investigated, designed, fabricated, and tested at Ku-band.

In summary, various challenges in the analysis and design of transmitarray antennas are addressed in this book. New methodologies to improve the bandwidth of transmitarray antennas have been demonstrated. Several prototypes have been fabricated and tested, demonstrating the desirable features and potential new applications of transmitarray antennas.

KEYWORDS

transmitarray antennas, frequency selective surfaces, multilayer aperture antennas, high gain antennas, wideband transmitarray antennas, multibeam transmitarray antennas

Dedicated to my parents, my beloved wife Heba,
and my children Farida and Adam.

– Ahmed H. Abdelrahman

Dedicated to my colleagues and family.

– Fan Yang

To my wife, Magda, daughters, Dalia and Donia,
son, Tamer, and the memory of my parents.

– Atef Z. Elsherbeni

To my parents.

– Payam Nayeri

Contents

List of Figures

List of Tables

Acknowledgments

The authors acknowledge the support from ANSYS and CST for providing the simulation software, HFSS and CST Microwave Studio, to use in many of our designs, and the support from Rogers for providing the substrate material for building the transmitarray antennas.

We also acknowledge the NSF support for this research project under contract # ECCS-1413863.

Ahmed H. Abdelrahman, Fan Yang, Atef Z. Elsherbeni, and Payam Nayeri
January 2017

CHAPTER 1

Introduction

1.1 TRANSMITARRAY ANTENNA CONCEPT

Operating based on the principles of electromagnetics, antennas are important electronic devices that are used in a wide range of applications such as broadcasting, radar, wireless communications, remote sensing, and space exploration. Although antennas have a history of over 100 years, new antenna concepts keep on emerging because of the exploration of new frequency spectrum such as THz band, advancements on materials, and fabrication techniques, as well as increasing computational and experimental capabilities. Transmitarray antenna is such a transformative and exciting concept that attracts growing interests of many researchers in the antenna area.

The vast diversities of antennas can be classified into low gain antennas (<10 dBi), middle gain antennas (10–20 dBi), and high gain antennas (>20 dBi). Transmitarray antennas belong to the high gain antenna group. Traditionally, a high gain can be realized using two approaches: one is based on the optic theory that manipulates the geometrical curvature of antenna surface to focus the radiation beam; the other is the antenna array theory that controls the interference of elements radiation appropriately. Representations for the first approach are the parabolic reflectors and lens antennas, and examples of the second approach include waveguide-slot arrays and printed microstrip antenna arrays. As an emerging concept, the transmitarray antenna combines the favorable features of optic theory and antenna array techniques, leading to a low profile conformal design with high radiation efficiency and versatile radiation performance.

A transmitarray antenna consists of an illuminating feed source and a thin transmitting surface, as shown in Fig. 1.1. The feed source is located on an equivalent focal point. On the transmitting surface, there is an array of antenna elements. The transmission coefficients of these elements are individually designed to convert the spherical phase front from the feed to a planar phase front. As a result, a focused radiation beam can be achieved with a high gain.

Transmitarray antennas have a great potential in many applications such as earth remote sensing, wireless communications, spatial power combining for high power applications, THz images and sensors, and solar energy concentrator.

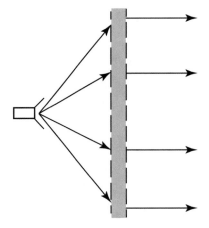

Figure 1.1: Geometry of a transmitarray antenna.

1.2   COMPARISON WITH SOME RELATED ANTENNA TECHNOLOGIES

It is realized that there exist some related technologies, both from microwaves and from optics. Some of these antennas are illustrated in Fig. 1.2, and their relations with transmitarray antennas are explained one by one.

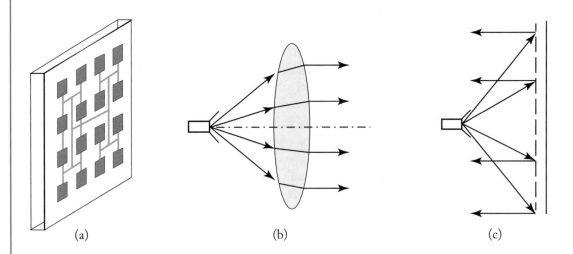

Figure 1.2: Antenna technologies related to transmitarrays: (a) a planar phased array, (b) a lens antenna, and (c) a planar reflectarray.

Both planar phased array antenna, shown in Fig. 1.2a, and transmitarray have the capability to individually control the element excitation to generate a focused beam. A major difference is the feed mechanism: the planar phased array uses a feeding network whereas the transmitarray uses a space feeding source. Despites the design complexity, the feeding network suffers from a severe energy loss, which impedes its implementation in large-aperture or high-frequency applications, such as THz exploration. In contrast, the transmitarray is more energy efficient due to the space feeding scheme.

Lens antenna, shown in Fig. 1.2b, uses the same space feeding as the transmitarray; thus, it is popularly used at high frequency all the way to optic range. However, the curved surface of the lens increases the fabrication complexity. In contrast, the planar transmitting surface can be fabricated using the standard low-cost printed circuit board (PCB) technique. As frequency increases and wavelength reduces, it can also be readily produced with micro- or nano-fabrication techniques. Therefore, the unified fabrication approach for transmitarrays and other circuit components enables an integrated system design and a reduced system cost. Furthermore, compared to a Fresnel lens that could also be built in a planar geometry, the transmitarray antenna is much more efficient because of the unique control of the transmission magnitude and phase.

Figure 1.2c shows the geometry of a planar reflectarray. The relation between a reflectarray and a transmitarray is similar to the relation between a mirror and a lens. Although inspired by the reflectarray, the transmitarray encounters a great challenge: both magnitude and phase control of the array element. In the reflectarray, the reflection magnitude is always 1 (0 dB) due to the existence of a metal ground plane that reflects the entire incident wave; thus, one only needs to control the element reflection phase. In transmitarray, besides the phase control, the magnitude of the transmission coefficient needs to be close to 1 (0 dB) to ensure a high efficiency.

1.3 TRANSMITARRAY DESIGN APPROACHES

There are different approaches to design transmitarray antennas. Among them, the representative design techniques are:

(a) multi-layer frequency selective surfaces (M-FSS);

(b) receiver-transmitter design; and

(c) metamaterial/transformation approach.

1.3.1 MULTI-LAYER FREQUENCY SELECTIVE SURFACES (M-FSS)

The array of printed elements on the transmitarray antenna surface aims to convert the spherical phase front from the antenna feed to a planar phase front. We can control the phase of each array element individually by varying its dimensions [1]–[6]. However, the phase compensation cannot be achieved by only one layer of frequency selective surface (FSS) [1, 2], while a multi-layer FSS structure separated by either air gap or thick substrate, as shown in Fig. 1.3, is required to increase the transmission phase range.

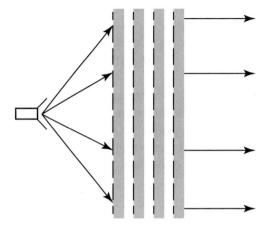

Figure 1.3: Multi-layer FSS configuration.

1.3.2   RECEIVER-TRANSMITTER DESIGN

A receiver-transmitter configuration typically consists of two planar arrays of printed antennas, whose elements are coupled or interconnected with transmission lines. The first array is illuminated by an antenna feed source, and it acts as a receiver. The coupling structures or transmission lines between two planar arrays are designed to achieve a specific phase and magnitude distribution from the first array to the second array, which acts as a transmitter radiating wave into free space [7]–[19]. A unit-cell element of the receiver-transmitter configuration is shown in Fig. 1.4.

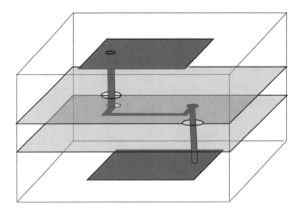

Figure 1.4: Receiver-transmitter configuration.

1.3.3   METAMATERIAL/TRANSFORMATION APPROACH

Another approach to control the element phase of the antenna array is to vary the effective substrate permittivity and permeability using metamaterial configuration [20]–[23]. In [20], a negative reflector index lens antenna using dielectric resonators was designed to achieve a wide beam scanning radiation pattern. A 2D broadband low-loss Luneburg lens was designed using complementary I-shaped unit-cell metamaterials, which is demonstrated in [21]. In [22], a new type of gradient-index metamaterial, composed of a dielectric post array, is proposed for millimeter-wave applications to achieve wideband, low-reflection characteristics, and low sensitivity to the polarization of the incident wave. A flat lens synthesis is carried out using a systemic phase-shifting strategy. In [23], a technique for designing true-time-delay microwave lenses with low-profile and ultra-wideband performances is proposed. The proposed lens is composed of numerous spatial true-time-delay units distributed over a planar surface. Each spatial true-time-delay unit is the unit-cell of an appropriately designed metamaterial structure, which is composed entirely of non-resonant constituting elements.

1.4   OVERVIEW OF RESEARCH TOPICS

The goal of this book is the study of the transmitarray antenna design, for being one of most prominent types of high gain antennas. Transmitarray antenna has received considerable attention in recent years, and it carries a lot of challenges to achieve better performance in various applications. Figure 1.5 summarizes the content of this book.

Chapter 2 introduces the main equations required to calculate the radiation pattern, directivity, and gain of the planar space-fed array antennas such as reflectarrays and transmitarrays. It presents three different directivity calculation methods and explains how different their results are from each other. It also discusses phase error analysis, explaining the different sources of phase errors, and their effects to the antenna design.

Chapter 3 presents an analytical analysis of the transmission coefficient of multi-layer conductors separated by dielectric material for transmitarray designs. It investigates the transmission behaviors and reveals the transmission phase limit of the multi-layer frequency selective surfaces (M-FSS) configuration, which will be general for arbitrary FSS geometries. The effectiveness of the analytical study has been validated through numerical simulations of several representative FSS examples.

Chapter 4 presents detailed design analysis of a multiple conductor layers transmitarray antenna using a new element of slot-type. This design has a novelty in using slot-type element with no dielectric substrate, which has the advantages of low cost and suitability for space applications. The impact of the element shape on the overall gain and radiation pattern is discussed, taking into account the oblique incidence angles and the feed polarization conditions.

Chapter 5 aims to reduce the complexity and cost of transmitarray antennas by decreasing the number of layers. It demonstrates three different methods to design triple-layer transmitarray antennas, while maintaining the overall performance with full 360° transmission phase of the

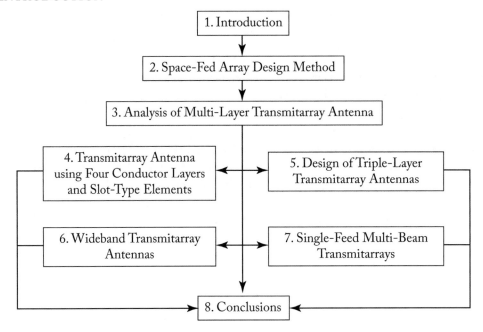

Figure 1.5: Content of the book.

transmitarray antenna. Based on this study, a high gain prototype transmitarray antenna using spiral type elements is designed, fabricated, and tested.

Chapter 6 discusses the frequency variations in the transmission phase and magnitude of the unit-cell element, and demonstrates the transmitarray bandwidth vs. the transmission phase range of the unit-cell element. Furthermore, the effect of the reference phase on the performance of the transmitarray antenna is discussed. Comparisons between three different element shapes are also considered. Finally, two different transmitarray antennas are designed, fabricated, and tested at Ku-band to demonstrate the bandwidth performance.

Chapter 7 discusses the radiation characteristics of single-feed transmitarray antennas with simultaneous multiple beams, through case studies of quad-beam designs. Various pattern masks and fitness functions are studied for multi-beam designs. A Ku-band quad-beam transmitarray antenna is successfully demonstrated.

Chapter 8 provides the conclusion of this research.

CHAPTER 2

Space-fed Array Design Method

This chapter presents detailed analysis of the design of planar space-fed array antennas such as reflectarrays and transmitarrays, with more focus on the requirements of transmitarray antennas. First the basics of the aperture phase distribution and the analysis of the array elements are described. Next the radiation performances of the space-fed arrays are described using the array theory approach. Three different methods to calculate the directivity of the space-fed arrays are then presented with comparisons of the result accuracies and computational times. Antenna gain calculations are presented next when taking into account the antenna spillover and the element losses. Finally, discussed is an extensive study of element phase errors, with clarification of the error origins, as well as the phase error impact on the radiation pattern and gain of the antenna.

2.1 PHASE DISTRIBUTION ON TRANSMITARRAY APERTURE

The analysis of a transmitarray antenna starts with an assumption that the transmitarray elements are in the far-field region of the feed source, which is usually located in a centered position. In this case, the electromagnetic field incident on each transmitarray element at a certain angle can be locally considered as a plane wave with a phase proportional to the distance from the phase center of the feed source to each element, as corresponds to spherical wave propagation.

The required transmission phase of each transmitarray element is designed to compensate the spatial phase delay from the feed horn to that element, so that a certain phase distribution can be realized to focus the beam at a specific direction, as shown in Fig. 2.1. The transmission phase ψ_i for the ith element is calculated as [24, 25]:

$$\psi_i = k\left(R_i - \vec{r}_i \cdot \hat{r}_o\right) + \psi_0, \tag{2.1}$$

where k is the propagation constant in free space, R_i is the distance from the feed horn to the ith element, $\vec{r}_i$ is the position vector of the ith element, and the main beam direction is represented by $\hat{r}_o$. For a transmitarray with a main beam at the broadside direction, $\vec{r}_i \cdot \hat{r}_o = 0$. The parameter ψ_0 is a constant phase, indicating that a relative transmission phase rather than the absolute transmission phase is required for transmitarray design. Equation (2.1) is general for space-fed array design, e.g., reflectarray and transmitarray antennas.

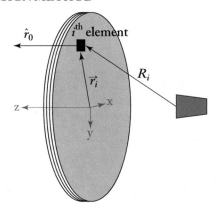

Figure 2.1: Phase compensation of a multi-layer transmittarray antenna.

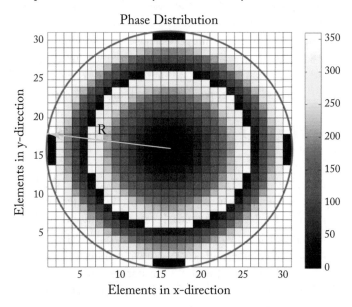

Figure 2.2: Example of the required phase distribution in a circular aperture transmittarray.

Figure 2.2 presents the required phase distribution of a circular aperture transmitarray antenna of 30×30 elements with half wavelength unit-cell periodicity, and focal length to diameter ratio $F/D = 0.8$. The focal point is centered, and a pencil beam is produced in the broadside direction. The circular boundary of the array aperture defined here is:

$$R = \left[\left(\frac{M}{2} \right) + 0.1 \right] P,$$

where R is the radius of the circular aperture, M is the number of unit-cell elements along the main axes, and P is the unit-cell periodicity. This definition includes a margin of $0.1P$. Hint: the circular boundary could differ from one designer to another.

Once the required transmission phase is determined for each element on the transmitarray aperture, the corresponding element dimension is obtained using the transmission phase vs. element dimension curve which is usually obtained from the unit-cell full EM wave analysis, while maintaining the element transmission magnitude close to 1 (0 dB).

2.2   UNIT-CELL ELEMENT ANALYSIS

A key feature of transmitarray implementation is how the individual elements are designed to transmit electromagnetic waves with the desired phases. Section 1.3 discusses three different techniques that control the transmission phases of the individual transmitarray elements. However, it is worthwhile to clarify how to analyze the element designs and determine the accurate characterization of the array elements.

To obtain the required transmission characteristics of the array elements, usually a parametric study of the unit-cell element is performed depending on the transmitarray design approach. The transmitarray element is usually simulated in a unit-cell, as shown in Fig. 2.3, with linked periodic boundary conditions [26], which mimics the periodic environment of the elements. Most electromagnetic simulators have the capability to analyze the transmission characteristics of the unit-cell element, such that the element transmission phase and the corresponding transmission magnitude can be obtained.

Figure 2.4 depicts the transmission magnitude and phase vs. the element dimensions of the unit-cell element of Fig. 2.3. The detailed dimensions of the unit cell are provided in Section 6.1, and the operation frequency is 13.5 GHz. It is observed that when the loop length changes, a 360° phase change is achieved while the transmission magnitude is close to 0 dB. The simulations of the unit-cell element are performed under certain approximations. These approximations are presented in Section 2.6.2.

2.3   RADIATION ANALYSIS USING THE ARRAY THEORY

Far-field radiation pattern of a space-fed array antenna, e.g., reflectarray and transmitarray, can be calculated using the conventional array theory. The radiation pattern of a 2D planar array with $M \times N$ elements can be calculated as [25, 28]:

$$\vec{E}\left(\hat{u}\right) = \sum_{m=1}^{M} \sum_{n=1}^{N} \vec{A}_{mn}\left(\hat{u}\right) \cdot \vec{I}\left(\vec{r}_{mn}\right),$$

$$\hat{u} = \hat{x} \sin\theta \cos\varphi + \hat{y} \sin\theta \sin\varphi + \hat{z} \cos\theta,$$

(2.2)

where $\vec{A}$ is the element pattern vector function, $\vec{I}$ is the element excitation vector function, and $\vec{r}_{mn}$ is the position vector of the mn^{th} element.

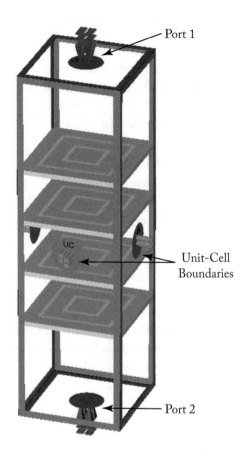

Port 1

Unit-Cell Boundaries

Port 2

Figure 2.3: A 3D model a quad-layer transmitarray unit-cell in CST Microwave Studio software [27].

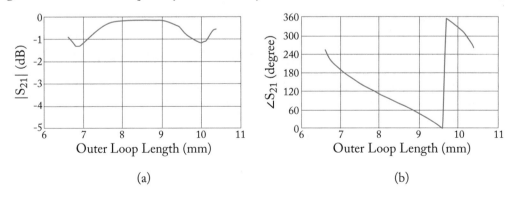

(a)

(b)

Figure 2.4: Transmission coefficient of a quad-layer unit-cell at 13.5 GHz.

A general coordinate system of the transmitarray antenna is given in Fig. 2.5. The origin of the coordinate system is located at the center of the aperture, and the x and y axes are set on the aperture plane. The aperture plane is illuminated by a feed source located at height H from the aperture and has a projection on the y-axis. Therefore, the feed has coordinates $(0, -H \tan \alpha, -H)$, where α is the offset angle.

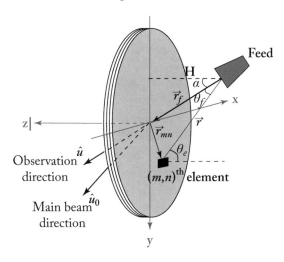

Figure 2.5: The coordinate system of the transmitarray antenna.

To simplify calculations, scalar functions are usually used in the analysis. For the element-pattern function A, a cosine q model is considered for each element with no azimuthal dependence [25, 28], i.e.,

$$A_{mn}(\theta, \varphi) \approx \cos^{q_e}(\theta) e^{jk(\vec{r}_{mn} \cdot \hat{u})}, \qquad (2.3)$$

where q_e is the element pattern power factor. The illumination of the aperture can be obtained by using another cosine q model as an approximation of the feed horn pattern, and taking into account the Euclidian distance between the feed horn and the element. The element excitation can then be expressed as [25, 28]:

$$I(m,n) \approx \frac{\cos^{q_f}(\theta_f(m,n))}{|\vec{r}_{mn} - \vec{r}_f|} \cdot e^{-jk|\vec{r}_{mn} - \vec{r}_f|} \cdot |T_{mn}| e^{j\psi_{mn}}, \qquad (2.4)$$

where $\theta_f(m,n)$ is the spherical angle in the feed's coordinate system, q_f is the feed pattern power factor, $\theta_e(m,n)$ is the angle between the line from feed to the mn^{th} element $|\vec{r}|$ and the normal direction of the aperture plane, $\vec{r}_f$ is the position vector of the feed, $|T_{mn}|$ is the transmission magnitude of the mn^{th} element which is obtained directly from the unit-cell analysis, and ψ_{mn} is the required phase delay of the mn^{th} element to set the main beam in the $\hat{u}_o$ direction, as described in Equation (2.1).

The mutual coupling effect between the transmitarray elements is considered during the element analysis using the infinite array approach represented by the use of the periodic boundary condition. This approach analyzes each transmitarray element in an infinite array environment, with all surrounding elements considered identical.

With these approximations, the radiation pattern based on Equation (2.2) can be simplified to the scalar form [25, 28]:

$$E(\theta, \varphi) = \sum_{m=1}^{M} \sum_{n=1}^{N} \cos^{q_e}(\theta) \frac{\cos^{q_f}(\theta_f(m,n))}{|\vec{r}_{mn} - \vec{r}_f|}$$
$$\cdot e^{-jk(|\vec{r}_{mn} - \vec{r}_f| - \vec{r}_{mn} \cdot \hat{u})} \cdot |T_{mn}| e^{j\psi_{mn}}. \tag{2.5}$$

Equation (2.5) calculates the radiation pattern of a rectangular aperture transmitarray of size $M \times N$ elements. For circular aperture transmitarray, Equation (2.5) can still be used but with ignoring those elements that are outside of the circular aperture ($|\vec{r}_{mn}| >$ aperture radius).

Figure 2.6 presents the radiation pattern of a centered-fed circular aperture transmitarray antenna of 30×30 elements with half wavelength unit-cell periodicity. The focal length to diameter ratio $F/D = 0.8$, the feed pattern power factor $q_f = 6$, and the element pattern power factor $q_e = 1$. Although F/D parameter is not directly presented in Equation (2.5), its importance is demonstrated in the phase distribution of the array elements ψ_{mn} and the elements' excitation $I(m,n)$ on the array aperture.

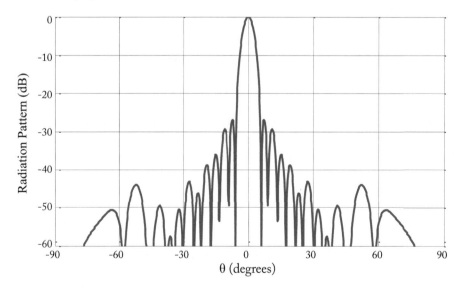

Figure 2.6: Radiation pattern of a 30×30 circular apperture transmitarray antenna with a broadside beam.

2.4 DIRECTIVITY CALCULATIONS

The directivity of an antenna is the ratio of the maximum radiation intensity to the average radiation intensity over all directions. In mathematical form, it can be written as [29]:

$$D_0 = \frac{U_{max}}{U_0} = \frac{U_{max}}{\frac{1}{4\pi} P_{rad}}, \tag{2.6}$$

where U_{max} is the maximum radiation intensity, U_0 is the radiation intensity of an isotropic source, and P_{rad} is the total radiated power.

Once the radiation pattern of the transmitarray antenna is obtained using Equation (2.5), the antenna directivity can then be obtained as:

$$D_0 = \frac{|E(\theta_0, \varphi_0)|^2}{\frac{1}{4\pi} \int_0^{2\pi} \int_0^\pi |E(\theta, \varphi)|^2 \sin\theta \, d\theta d\varphi}, \tag{2.7}$$

where θ_0 and φ_0 are the direction of the main beam. The calculated directivity in Equation (2.7) takes into consideration the illumination (taper) efficiency of the array, and the effect of the projected aperture for off-broadside beams [25]. The main challenge in solving Equation (2.7) is the evaluation of the denominator, which is done numerically. This section presents three different methods to calculate the space-fed array directivity, and clarifies the approximations and the required computational time of each method.

2.4.1 METHOD 1: NUMERICAL INTEGRATION

This method is simply calculates the double integration in the denominator of Equation (2.7) as double summation over the angles θ and φ such that [30]:

$$\int_0^{2\pi} \int_0^\pi |E(\theta, \varphi)|^2 \sin\theta \, d\theta d\varphi \approx \sum^{N_\theta} \sum^{N_\varphi} |E(\theta, \varphi)|^2 \sin\theta \, \Delta\theta \Delta\varphi, \tag{2.8}$$

where $\Delta\theta$ and $\Delta\varphi$ are the unit step in θ and φ directions, N_θ and N_φ are the number of integration steps. The directivity is then given by:

$$D_0 = \frac{|E(\theta_0, \varphi_0)|^2}{\frac{1}{4\pi} \sum^{N_\theta} \sum^{N_\varphi} |E(\theta, \varphi)|^2 \sin\theta \, \Delta\theta \Delta\varphi}. \tag{2.9}$$

By considering the double summation of Equation (2.5), the computational time of the denominator in Equation (2.7) is an $O(M \times N \times N_\theta \times N_\varphi)$. As the unit steps $\Delta\theta$ and $\Delta\varphi$ decrease (N_θ and N_φ increase), the accuracy of the directivity calculation improves, but at the expenses of the computational time. Figure 2.7 depicts the directivity calculations vs. $\Delta\theta$ and $\Delta\varphi$ for the transmitarray antenna example that has phase distribution and radiation pattern shown in Figs. 2.2 and 2.6, respectively.

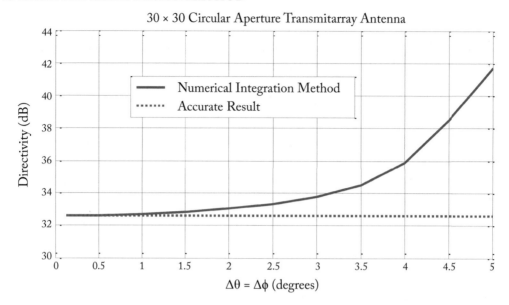

Figure 2.7: Directivity calculations of a circular aperture transmitarray antenna using the numerical integration method.

It is worthwhile to evaluate the computational time of the directivity calculations vs. $\Delta\theta$ and $\Delta\varphi$. Table 2.1 depicts values of the directivity shown in Fig. 2.7 and the corresponding computational time at different $\Delta\theta$ and $\Delta\varphi$ values. These calculations are done on a 2.66 GHz Intel(R) Core(TM)2 CPU computer with 4 GB of memory using Matlab version R2011b.

It is clear that the main disadvantage of the numerical integration method for directivity calculations is the high computational time for obtaining accurate results.

2.4.2   METHOD 2: UTILIZATION OF BESSEL FUNCTION

The radiation pattern of Equation (2.5) can also be written as [30]:

$$E(\theta,\varphi) = \sum_{i=1}^{T} w_i e^{jk(\vec{r}_i \cdot \hat{u})}, \tag{2.10}$$

where T is the total number of elements ($T = N \times M$), and

$$w_i = \frac{\cos^{q_f}\left(\theta_f(m,n)\right)}{|\vec{r}_{mn} - \vec{r}_f|} \cdot e^{-jk(|\vec{r}_{mn} - \vec{r}_f|)} \cdot |T_{mn}|e^{j\psi_{mn}}. \tag{2.11}$$

Table 2.1: Directivity calculations and the corresponding computational time of a 30 × 30 circular aperture transmitarray antenna with different $\Delta\theta$ and $\Delta\phi$ values

$\Delta\theta = \Delta\varphi$ (degrees)	Directivity (dB)	Computation Time (seconds)
5.0	41.70	0.71
4.5	38.53	0.85
4.0	35.90	1.05
3.5	34.51	1.35
3.0	33.77	1.89
2.5	33.33	2.67
2.0	33.03	4.12
1.5	32.83	7.25
1.0	32.69	16.15
0.75	32.65	28.42
0.50	32.61	64.22
0.25	32.60	266.46
0.125	32.59	1060.01

Here we assume isotropic radiation from each element ($q_e = 0$). The term $\vec{r}_i \cdot \hat{u}$ in Equation (2.10) can be written as:

$$\begin{aligned}
\vec{r}_i \cdot \hat{u} &= \left(\hat{x} P_{x_i} + \hat{y} P_{y_i} + \hat{z} 0\right) \cdot \left(\hat{x} \sin\theta \cos\varphi + \hat{y} \sin\theta \sin\varphi + \hat{z} \cos\theta\right) \\
&= P_{x_i} \sin\theta \cos\varphi + P_{y_i} \sin\theta \sin\varphi,
\end{aligned} \tag{2.12}$$

where $\vec{r}_i = \left(P_{x_i}, P_{y_i}, 0\right)$ is the position vector of the ith element. Equation (2.10) can then be written as [30]:

$$E(\theta, \varphi) = \sum_{i=1}^{T} w_i e^{jk\left(P_{x_i} \sin\theta \cos\varphi + P_{y_i} \sin\theta \sin\varphi\right)}. \tag{2.13}$$

The denominator of Equation (2.7) can be written as:

$$\begin{aligned}
DEN &= \frac{1}{4\pi} \int_0^{2\pi} \int_0^{\pi} |E(\theta, \varphi)|^2 \sin\theta \, d\theta d\varphi \\
DEN &= \sum_{i=1}^{T} \sum_{j=1}^{T} w_i w_j^* \int_0^{\pi} \frac{1}{2} \sin\theta \, d\theta \int_0^{2\pi} \\
&\qquad \frac{1}{2\pi} e^{jk\left(\Delta P_{x_{ij}} \sin\theta \cos\varphi + \Delta P_{y_{ij}} \sin\theta \sin\varphi\right)} d\varphi,
\end{aligned} \tag{2.14}$$

where

$$\Delta P_{x_{ij}} \triangleq P_{x_i} - P_{x_j} \quad \text{and} \quad \Delta P_{y_{ij}} \triangleq P_{y_i} - P_{y_j}. \tag{2.15}$$

Now define

$$\rho_{ij} = \sqrt{\left(\Delta P_{x_{ij}}\right)^2 + \left(\Delta P_{y_{ij}}\right)^2} = \left\| \Delta P_{x_{ij}} \right\|, \tag{2.16}$$

$$\gamma_{ij} = \tan^{-1}\left(\frac{\Delta P_{y_{ij}}}{\Delta P_{x_{ij}}}\right). \tag{2.17}$$

Then Equation (2.15) can be written as:

$$\Delta P_{x_{ij}} = \rho_{ij} \cos \gamma_{ij} \quad \text{and} \quad \Delta P_{y_{ij}} = \rho_{ij} \sin \gamma_{ij}. \tag{2.18}$$

Using Equation (2.18), the exponential term in the inner integral in Equation (2.14) is

$$jk \left(\Delta P_{x_{ij}} \sin \theta \cos \varphi + \Delta P_{y_{ij}} \sin \theta \sin \varphi\right)$$
$$= jk \left(\rho_{ij} \sin \theta \cos \varphi \cos \gamma_{ij} + \rho_{ij} \sin \theta \sin \varphi \sin \gamma_{ij}\right)$$
$$= jk\rho_{ij} \sin \theta \cos\left(\varphi - \gamma_{ij}\right). \tag{2.19}$$

Using Equation (2.19) in the inner integral in Equation (2.14) gives

$$\int_0^{2\pi} \frac{1}{2\pi} e^{jk\left(\Delta P_{x_{ij}} \sin \theta \cos \varphi + \Delta P_{y_{ij}} \sin \theta \sin \varphi\right)} d\varphi$$
$$= \int_0^{2\pi} \frac{1}{2\pi} e^{jk\rho_{ij} \sin \theta \cos\left(\varphi - \gamma_{ij}\right)} d\varphi,$$
$$= J_0 \left(k \rho_{ij} \sin \theta\right), \tag{2.20}$$

where $J_0(\cdot)$ is the Bessel function of order zero.
Substituting Equation (2.20) into Equation (2.14) gives

$$DEN = \sum_{i=1}^{T} \sum_{j=1}^{T} w_i w_j^* \int_0^{\pi} \frac{1}{2} \sin \theta \, J_0 \left(k \rho_{ij} \sin \theta\right) d\theta, \tag{2.21}$$

$$DEN = \sum_{i=1}^{T} \sum_{j=1}^{T} w_i w_j^* sinc(k\rho_{ij}). \tag{2.22}$$

where the cardinal sine function or sinc function is defined as:

$$sinc(x) = \frac{\sin(x)}{x}. \tag{2.23}$$

The directivity is then given by:

$$D_0 = \frac{|E(\theta_0, \varphi_0)|^2}{\sum_{i=1}^{T} \sum_{j=1}^{T} w_i w_j^* sinc(k\rho_{ij})}. \tag{2.24}$$

It is worthwhile to notice that Equation (2.24) is independent of θ and φ. The computational time of Equation (2.24) is an $O(T^2)$.

For the same example of 30×30 transmitarray antenna of Table 2.1 with the utilization of Bessel function method, the calculated directivity equal to 32.59 dB and the computational time is 0.27 s. The same directivity value is obtained using numerical integration of method 1 with computational time of 1060.01 s, as shown in Table 2.1.

2.4.3 METHOD 3: ILLUMINATION EFFICIENCY

The directivity of a space-fed array can also be calculated in terms of its own aperture radiation and the illumination on the aperture, such that [31, 32]:

$$D_0 = D_{\max} \cdot \eta_{ill}, \tag{2.25}$$

where $D_{\max}$ is the maximum directivity of the aperture, which is function of the aperture area (A_p) and the operating frequency:

$$D_{\max} = \frac{4\pi}{\lambda^2} A_p. \tag{2.26}$$

η_{ill} is the illumination efficiency, which is the efficiency loss due to the non-uniform amplitude and phase distribution on the aperture plane and is equal to the product of the taper efficiency and the phase efficiency as follows:

$$\eta_{ill} = \eta_{taper} \cdot \eta_{phase}. \tag{2.27}$$

The taper efficiency η_{taper} accounts for the aperture illumination taper due to the feed and the antenna geometry and is given by [31]:

$$\eta_{taper} = \frac{1}{A_p} \frac{\left[\int_s |I_i|\, dS\right]^2}{\int_s |I_i|^2 dS} = \frac{1}{A_p} \frac{\left[\sum_{i=1}^{T} |I_i|\, \Delta x \Delta y\right]^2}{\sum_{i=1}^{T} |I_i|^2 \Delta x \Delta y}, \tag{2.28}$$

where I_i is the aperture field, dS is the unit area. For the ith element,

$$I_i = \cos^{q_e}(\theta) \frac{\cos^{q_f}\left(\theta_f(m,n)\right)}{|\vec{r}_i - \vec{r}_f|} \cdot e^{-jk(|\vec{r}_i - \vec{r}_f| - \vec{r}_i \cdot \hat{u})} \cdot |T_{mn}| e^{j\psi_i}. \tag{2.29}$$

The phase efficiency (η_{phase}) accounts for the phase error over the aperture and is given by [31]:

$$\eta_{phase} = \frac{\left[\int_s I_i dS\right]^2}{\left[\int_s |I_i| dS\right]^2} = \frac{\left|\sum_{i=1}^{T} I_i \Delta x \Delta y\right|^2}{\left[\sum_{i=1}^{T} |I_i| \Delta x \Delta y\right]^2}. \tag{2.30}$$

Substituting Equations (2.28) and (2.30) in Equation (2.27), the illumination efficiency is given by:

$$\eta_{ill} = \frac{1}{A_p} \frac{\left[\int_s I_i dS\right]^2}{\int_s |I_i|^2 dS} = \frac{1}{A_p} \frac{\left|\sum_{i=1}^T I_i \Delta x \Delta y\right|^2}{\sum_{i=1}^T |I_i|^2 \Delta x \Delta y}. \tag{2.31}$$

Substituting Equation (2.31) in Equation (2.25), the directivity is given by:

$$D_0 = \frac{4\pi}{\lambda^2} \frac{\left|\sum_{i=1}^T I_i \Delta x \Delta y\right|^2}{\sum_{i=1}^T |I_i|^2 \Delta x \Delta y}. \tag{2.32}$$

It is valuable to notice that Equation (2.32) is also independent of θ and φ and the computational time of this equation is an $O(T)$, which is much faster than the previous two methods.

For the same example of 30×30 transmitarray antenna of Table 2.1, the calculated directivity using the aperture efficiency method equal to 32.58 dB and the computational time is less than 0.01 s.

2.4.4   COMPARISON BETWEEN THE THREE METHODS

The directivity equations and the order of computational time of the three methods are summarized in Table 2.2. The aperture efficiency method is the most efficient in terms of time compared to the other two methods. The numerical integration method is more time consuming and the accuracy of its results depend on the number of integration steps N_θ and N_φ.

Table 2.3 depicts the directivity calculations and the computational time, using the three methods of calculations, for different size rectangular aperture transmitarray antennas with centered focal point, $F/D = 0.8, q_e = 1, q_f = 6$, half wavelength unit-cell periodicity, and main beam in the broadside direction.

The directivity increases with the increase of the array size as shown in Table 2.3. The three methods of calculations give very close directivity results, however the computational time of the numerical integration method is very high compared to the other two methods especially with the increase of the array size. The aperture efficiency method is the most time efficient in directivity calculations.

2.4.5   DIRECTIVITY BANDWIDTH

The array directivity changes with frequency due to the differential spatial phase delay resulting from the different lengths from the feed to each point on the wave front of the radiation beam. This variation is mainly represented mathematically in terms of the propagation constant k using Equation (2.5), assuming that the feed pattern power factor q_f does not change with frequency. Figure 2.8 presents the directivity vs. frequency of a 60×60 rectangular aperture transmitarray antenna with half wavelength unit-cell periodicity, centered focal point of

Table 2.2: Comparison between the three methods of directivity calculations

	Directivity Equation	Order of Computational Time
Method 1: Numerical Integration	$D_0 = \dfrac{\lvert E(\theta_0, \varphi_0)\rvert^2}{\frac{1}{4\pi}\sum^{N\theta}\sum^{N\varphi}\lvert E(\theta, \varphi)\rvert^2 \sin\theta\Delta\theta\Delta\varphi}$	$O(T \times N_\theta \times N_\varphi)$
Method 2: Utilization of Bessel Function	$D_0 = \dfrac{\lvert E(\theta_0, \varphi_0)\rvert^2}{\sum_{i=1}^{T}\sum_{j=1}^{T} w_i w_j^* \operatorname{sinc}(kp_{ij})}$	$O(T^2)$
Method 3: Aperture Efficiency	$D_0 = \dfrac{4\pi\lvert\sum_{i=1}^{T} E_i(\theta, \varphi)\Delta x\Delta y\rvert^2}{\lambda^2\sum_{i=1}^{T}\lvert E_i(\theta, \varphi)\rvert^2 \Delta x\Delta y}$	$O(T)$

Table 2.3: Directivity vs. rectangular array size

Array Size	Method 1 ($\Delta\theta = \Delta\varphi = 0.25°$)		Method 2		Method 3	
	Directivity (dB)	Computational Time (sec)	Directivity (dB)	Computational Time (sec)	Directivity (dB)	Computational Time (sec)
10 × 10	23.99	36.5	23.99	0.02	24.05	0.006
20 × 20	30.04	106.6	30.03	0.07	30.06	0.008
30 × 30	33.57	219.5	33.56	0.25	33.58	0.013
40 × 40	36.08	379.0	36.07	0.72	36.08	0.014
50 × 50	38.03	394.66	38.00	1.61	38.02	0.020
60 × 60	39.62	482.11	39.59	3.16	39.60	0.022

$F/D = 0.8, q_f = 6, q_e = 1$, and main beam in the broadside direction. The figure shows a clear agreement between the results obtained using the three methods for directivity calculations.

2.5   ANTENNA GAIN

Although the gain of the antenna is closely related to the directivity, it is a measure that takes into account the overall antenna efficiency. Among these efficiency types are the illumination efficiency, the spillover efficiency, and the element losses. The effect of the illumination efficiency is already taken into account when one calculates the radiation pattern directivity, as discussed in the previous section.

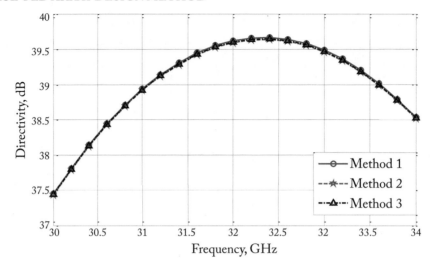

Figure 2.8: Directivity vs. frequency of 60×60 rectangular transmitarray antenna.

Once the spillover efficiency and the average element loss are determined, the transmitarray antenna gain can be calculated as:

$$G = D_0 \times \eta_{spill} \times EL_{avg},$$ (2.33)

where η_{spill} and EL_{avg} are the spillover efficiency and average element loss, respectively.

2.5.1 SPILLOVER EFFICIENCY

The spillover is the part of the power from the feed, which is not intercepted by the antenna aperture [33]. As illustrated in Fig. 2.9, the spillover efficiency is evaluated using the following Equation [31, 32]:

$$\eta_{spill} = \frac{\iint_\sigma \vec{P}(\vec{r}) \cdot d\vec{S}}{\iint_\Sigma \vec{P}(\vec{r}) \cdot d\vec{S}}.$$ (2.34)

The numerator represents the part of the power incident on the array aperture, and the denominator is the total power radiated by the feed. Both integrals are the fluxes of the pointing vector $\vec{P}(\vec{r})$ through some certain surface areas. The vector $\vec{r}$ is the position vector from the feed, as shown in Fig. 2.5. Usually, the integral of the denominator is performed over the entire spherical surface centered at the feed, denoted by Σ, as shown in Fig. 2.9. The integral of the numerator is evaluated over a portion σ of the sphere, where σ and the array aperture share the same solid angle with respect to the feed.

Because the aperture dimensions are known, it is necessary to determine the boundary of σ in terms of the spherical coordinate of the feed. The boundary of σ can be easily determined

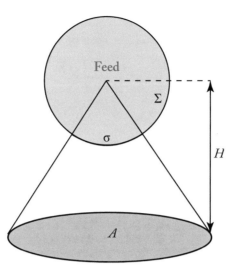

Figure 2.9: Spillover efficiency analysis.

for the case of having circular array aperture with a center feed. However, in general cases such as rectangular array aperture or an offset feed source, it is not so simple to determine the boundary of σ, and hence it is difficult to calculate the integral in the numerator of Equation (2.34). Accordingly, an alternative approach is to perform the integral over the array aperture A instead of on the surface σ of the sphere, because the array aperture and the spherical portion σ have the same solid angle with respect to the feed. Thus:

$$\eta_{spill} = \frac{\iint_A \vec{P}\left(\vec{r}\right) \cdot d\vec{S}}{\iint_\Sigma \vec{P}\left(\vec{r}\right) \cdot d\vec{S}}. \tag{2.35}$$

Equation (2.35) is general and straightforward. The integral in the numerator is calculated in a coordinate system that is suitable for the shape of the array boundary, as well as flexibility of having an arbitrary position of the feed.

To calculate the denominator of Equation (2.35), the pointing vector $\vec{P}\left(\vec{r}\right)$ should be defined in the spherical coordinates. The pointing vector of the feed defined by the cosine q model can be written in terms of the source region spherical coordinates as [32]:

$$\vec{P}\left(\vec{r}\right) = \hat{r}\frac{\cos^{2q_f}\left(\theta_f\right)}{r^2} \quad \left(0 \le \theta_f \le \frac{\pi}{2}\right), \tag{2.36}$$

where $r = \left|\vec{r}\right|$. Accordingly, the denominator of Equation (2.35) can be determined analytically as:

$$\iint_\Sigma \vec{P}\left(\vec{r}\right) \cdot d\vec{S} = \int_0^{2\pi}\int_0^{\pi/2} \cos^{2q_f}\left(\theta_f\right)\sin\left(\theta_f\right)d\theta_f d\varphi = \frac{2\pi}{2q_f + 1}. \tag{2.37}$$

To calculate the numerator of Equation (2.35), the pointing vector $\vec{P}(\vec{r})$ should be rewritten in the rectangular coordinates. Equation (2.36) can be written as:

$$\vec{P}(\vec{r}) = \hat{r}\frac{\cos^{2q_f}(\theta_f)}{r^2} = \vec{r}\frac{\cos^{2q_f}(\theta_f)}{r^3}$$

$$\vec{P}(\vec{r}) = [x\hat{x} + (y + H\tan\alpha)\hat{y} + (-H)\hat{z}]\frac{\cos^{2q_f}(\theta_f)}{r^3}, \tag{2.38}$$

where α is the feed offset angle as shown in Fig. 2.5. The case presented in Fig. 2.9 has no offset angle ($\alpha = 0$), and H is the focal length. Thus,

$$\vec{P}(\vec{r}) \cdot d\vec{S} = \frac{H\cos^{2q_f}(\theta_f)}{r^3} dx\, dy. \tag{2.39}$$

Accordingly, the numerator of Equation (2.35) can then be determined numerically as:

$$\iint_A \vec{P}(\vec{r}) \cdot d\vec{S} = \int_y \int_x \frac{H\cos^{2q_f}(\theta_f)}{r^3} dx\, dy$$
$$= \sum_{m=1}^{M}\sum_{n=1}^{N}\frac{H\cos^{2q_f}(\theta_f)}{r^3}\Delta x\,\Delta y, \tag{2.40}$$

where Δx and Δy are the unit-cell size along the x- and y-axes, respectively.

Substituting Equations (2.37) and (2.40) in Equation (2.35), the spillover efficiency can be determined as:

$$\eta_{spill} = \frac{2q_f + 1}{2\pi}\sum_{m=1}^{M}\sum_{n=1}^{N}\frac{H\cos^{2q_f}(\theta_f)}{r^3}\Delta x\,\Delta y. \tag{2.41}$$

2.5.2 ELEMENT LOSSES

The element losses include the conductor loss, the dielectric losses, and the reflection loss, which decrease the transmission magnitude of a transmitarray element. As mentioned in Section 2.2, the transmission characteristics of the unit-cell element, which include the element transmission phase and the corresponding transmission magnitude, can be obtained by using a full electromagnetic simulator. The conductor and dielectric losses can also be considered in the simulation process by selecting practical materials, which include the conductivity of the conductor layers and the loss tangent of the dielectric materials.

Once the transmission magnitude of each array element is determined, the average element loss can be calculated as:

$$EL_{avg} = \frac{\sum_{i=1}^{T} I_i^2 |T_i|^2}{\sum_{i=1}^{T} I_i^2}, \tag{2.42}$$

where I_i and $|T_i|$ are the illumination and the transmission coefficient magnitude of the ith element, respectively. T is the total number of elements ($T = N \times M$).

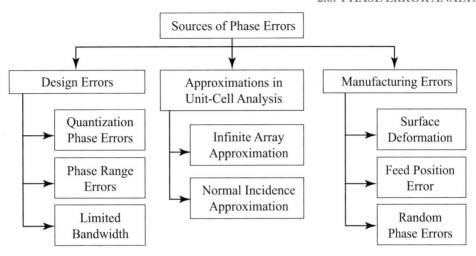

Figure 2.10: Sources of phase errors.

2.6   PHASE ERROR ANALYSIS

The transmission phases of the elements are key component in the transmitarray design. There-fore, it is worthwhile to figure out the sources of phase errors and study their impacts on the radiation performances. Element phase errors can be caused by design errors, approximations in the unit-cell analysis, and manufacturing errors, as shown in Fig. 2.10. A brief description of each of these errors is given in this section.

2.6.1   DESIGN ERRORS

These are errors that are produced during the practical design process. Among these errors are the quantization phase errors and the phase range errors. Furthermore, narrow band limitation of the unit-cell element and the differential spatial phase delay causes phase errors with the change of frequency, which limits the transmitarray bandwidth. The effects of these errors can be taken into account when calculating the radiation pattern and the gain of the transmitarray antenna. One just needs to replace the ideal element phase ψ_i of Equation (2.1) by the actual element phase during the radiation pattern and gain calculations. The actual element phase is obtained through the unit-cell element analysis that was discussed in Section 2.2 and based on the element dimensions.

Quantization Phase Errors
Section 2.1 explained how to calculate the required transmission phase of each transmitarray element using Equation (2.1). It also mentioned that the corresponding element dimension is obtained using the transmission phase vs. element dimension curve, which is obtained from the

unit-cell analysis discussed in Section 2.2. However, the manufacturing accuracy plays a significant role in the performance of the element. The element dimensions are changed by a certain amount depending on the manufacturing precision, and hence, a continuous phase control is not possible. Practically, the phase of each transmitarray element is selected to provide the closest quantization phase with respect to the ideal phase shift. The difference between the ideal element phase and the quantized phase of the selected element is categorized as quantization phase error [25].

Moreover, it is worth mentioning that there are some transmitarray antennas for certain applications use discrete tuning phase [9–11, 13, 15–19]. Despite the simplicity of this technique to design a specific number of unit-cells, it leads to the increase of the quantization phase errors. This subsection presents the impact of the quantization phase errors on both the radiation pattern and antenna gain. For simplicity, a constant quantization phase is assumed over the full phase range of 360°.

Using the example of 30 × 30 circular aperture transmitarray antenna that was presented in Section 2.1, the radiation patterns for different quantization phase values are demonstrated in Fig. 2.11, which shows increase in the side lobe level with the increase in the quantization phase. Figure 2.12 presents the transmitarray antenna gain vs. quantization phase, which illustrates reduction in antenna gain with the increase in the quantization phase.

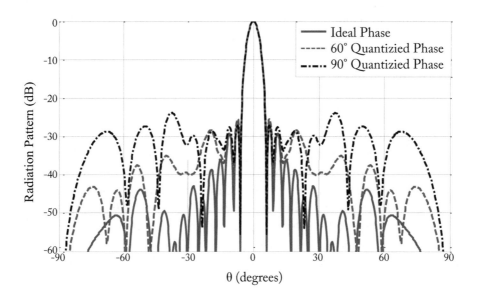

Figure 2.11: Radiation pattern of a circular aperture transmitarray antenna at different quantization phase values.

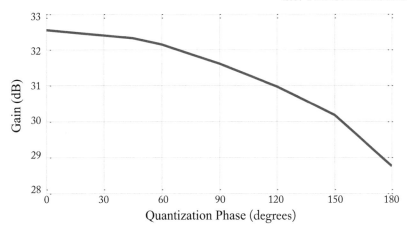

Figure 2.12: Transmitarray antenna gain vs. quantization phase.

Figure 2.13 presents the phase distributions of the transmitarray antenna at different quantization phase values comparing with the ideal phase distribution. The three cases that are shown in the figure are categorized as:

1. 3-bit or 8-state transmitarray antenna with the quantization phase equal to 45°;

2. 2-bit or 4-state transmitarray antenna with the quantization phase equal to 90°; and

3. 1-bit or 2-state transmitarray antenna with the quantization phase equal to 180°.

Phase Range Errors

The element phase range is one of the most important factors in element selection. Typically, a full transmission phase range of 360° cannot be achieved by only one layer of the printed antenna elements [1, 2], while multi-layer configuration is required to increase the transmission phase range. Chapter 3 studies in detail the multi-layer configuration, and reveals the corresponding transmission phase range that can be obtained. If the transmission phase range is smaller than the complete cycle of 360°, some elements will inevitably have unattainable phase shift. While the element selection routine minimizes the effect of these errors by selecting the closest quantized values, these errors are in nature different from the quantization errors and are categorized as errors due to limited phase range.

In order to clarify the impact of the limited phase range on the performances of the transmitarray antenna, we assume continuous phase (rather than practical quantized phase) in the available phase range. Using the same example of 30 × 30 circular aperture transmitarray antenna that was presented in Section 2.1, the radiation patterns for different element phase ranges are depicted in Fig. 2.14, which shows increase in the side lobe level with the decrease in the element

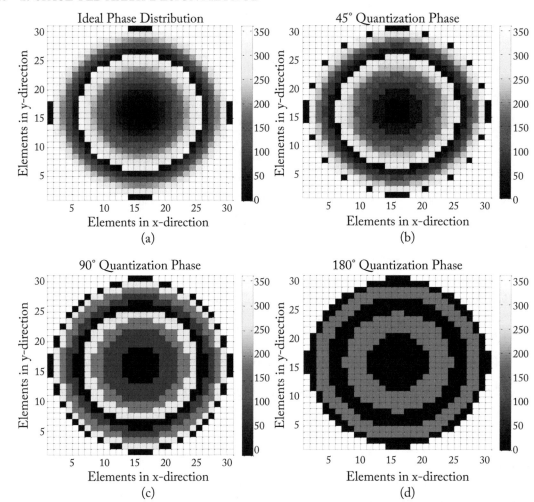

Figure 2.13: Phase distribution of the transmitarray antenna for different quantization phase values: (a) ideal phase distribution, (b) 3-bit phase distribution, (c) 2-bit phase distribution, and (d) 1-bit phase distribution.

phase range. Figure 2.15 presents the transmitarray antenna gain vs. element phase range, which illustrates the reduction in antenna gain due to the decrease in the element phase range.

Limited Bandwidth

The directivity bandwidth of the transmitarray antenna was discussed in Section 2.4.5. It was mentioned that the differential spatial phase delay, which is represented mathematically by the

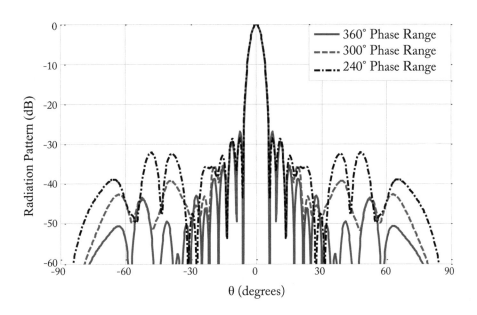

Figure 2.14: Radiation pattern of a circular aperture transmitarray antenna at different transmission phase ranges.

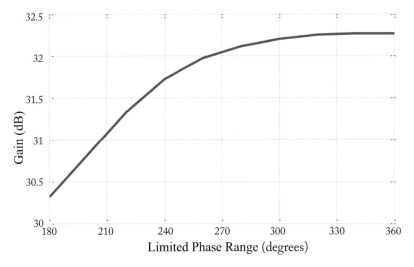

Figure 2.15: Transmitarray antenna gain vs. limited phase range.

propagation constant k, is a factor that limits the directivity bandwidth. In practice, the actual bandwidth of the transmitarray antenna is mainly affected by the narrow band limitation of the microstrip transmitarray elements. A certain element dimension has different corresponding transmission phases at different frequencies. Thus, the change of frequency leads to phase errors. Chapter 6 discusses in detail the factors that limit the transmitarray bandwidth, and presents how to handle these factors in order to increase the antenna bandwidth.

2.6.2 APPROXIMATIONS IN UNIT-CELL ANALYSIS

Since practical element situations in a transmitarray antenna system is complex, the transmission coefficient of the transmitarray elements is calculated by some degree of approximations. Among these approximations are the infinite array approximation and the normal incidence approximation. These approximations will introduce some phase errors and degrade the transmitarray performance.

Infinite Array Approximation

This approximation assumes that the transmission of an individual element surrounded by elements of different sizes can be approximated by a transmission from an infinite array of equal element size. This approximation is quite acceptable if the element dimensions don't vary significantly between adjacent elements. Some further discussions will be presented in Chapter 7 with a practical design example.

Normal Incidence Approximation

This approximation is to assume normal incidence for the element analysis, although the transmitarray antennas are generally illuminated by a spherical electromagnetic waves as demonstrated in Fig. 1.1. It has the advantage of simplifying the transmitarray design process, and works well with elements that have stable angular performance. However, if the incident angle is large or array elements are sensitive to the incident angle, it will cause notable phase errors. Depending on the design and the element shape, it might be necessary to model the element excitation angle and polarization accurately, and more details will be demonstrated in Chapter 4.

2.6.3 MANUFACTURING ERRORS

These are errors that arise from various causes such as manufacturing tolerances in the flatness of the array surface and etching process of the array elements [34], or displacement of the feed horn from the on-axis focus [30].

Surface Deformation

A space-fed array antenna, e.g., reflectarray or transmitarray antenna, illuminated by a spherical-wave feed is shown in Fig. 2.16. When a small surface deformation of height δ is introduced, the path length of a beam traveling from the feed to the antenna aperture changes from the correct path OA to the deformed path OB. This produces path length errors given approximately by:

$$\Delta = OA - OB \approx \delta \left(\frac{1}{\cos \theta} + 1 \right), \quad \text{for reflectarrays} \tag{2.43}$$

$$\Delta = OA - OB \approx \delta \left(\frac{1}{\cos \theta} - 1 \right), \quad \text{for transmitarrays.} \tag{2.44}$$

This Δ introduces a phase error, which deviates the aperture wavefront from the desirable plane wave, and leads to propagation gain reduction. From Equations (2.43) and (2.44), for identical θ and δ values, the transmitarray has less surface deformation error than that of the reflectarray.

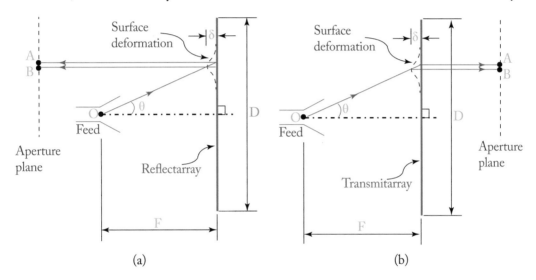

Figure 2.16: Surface deformation on: (a) reflectarray surface and (b) transmitarray surface.

Feed Position Error

The impact of a displacement of the feed horn from the on-axis focus is shown in Fig. 2.17. Assume that the field distribution has progressive phase over the aperture, and then the phase due to off-axis feed is linear in the coordinates over the aperture and causes an undistorted beam shift leading to a change in the direction of the main beam. The actual antenna gain in the z-direction (G_z) compared to the calculated gain (G) can then be given by [30]:

$$G_z = G \cos \theta_0, \tag{2.45}$$

$$\theta_0 \approx \theta_{off},$$

where θ_{off} is the feed offset angle from the on-axis focus, and θ_0 is the angle of main beam direction with respect to the broadside direction (z-direction).

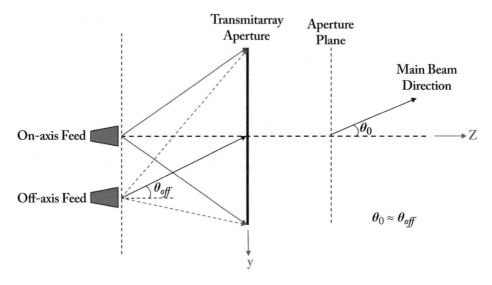

Figure 2.17: Feed position error of a transmitarray antenna.

Random Phase Errors

Random phase errors can be introduced by manufacturing tolerances due to surface deformation or during the etching process of the elements. The latter is usually more critical, which depends on the phase vs. element change curve. The antenna designer should minimize the slope of the phase vs. element change curve so that the change in phase is less sensitive to the change in element dimensions. If the curve is too steep, the element dimensions change due to manufacturing tolerances may become an issue, in particular at high microwave frequencies [24].

Random phase errors can be represented by adding a random phase value that has mean ($\mu = 0$) and standard deviation (σ) to the actual element phase. Figure 2.18 shows the bell curve of the standard normal distribution function with mean $\mu = 0$ and standard deviation $\sigma = 1$. In order to evaluate the effect of the random errors on the reduction of transmitarray antenna gain, the average of multiple trials has to be taken into account. Figure 2.19 shows six trials for transmitarray gain calculations with random phase errors vs. standard deviation of the random function. It shows reduction in the transmitarray antenna gain with the increase of the random phase errors.

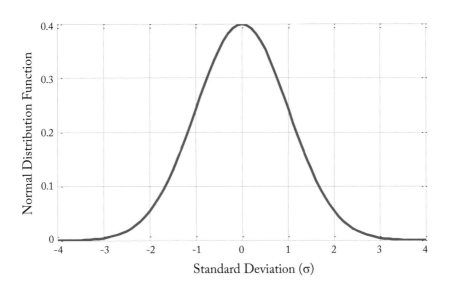

Figure 2.18: Standard normal distribution function.

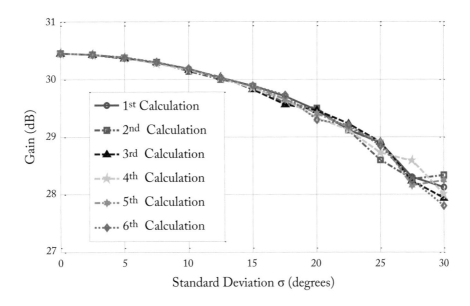

Figure 2.19: Effect of random phase error on the gain of a transmitarray antenna.

CHAPTER 3

Analysis of Multi-layer Transmitarray Antenna

Many transmitarray antennas are designed with multi-layer frequency selective surface (M-FSS) type elements. This approach is popularly used to control the transmission magnitude and phase of each element in the array individually by varying the element's dimensions [1]–[6]. However, a full 360° transmission phase range compensation cannot be achieved by only one layer of the printed antenna elements array [1, 2]. Thus, multi-layer design in which the layers are separated by either air gaps or dielectric material is required to increase the transmission phase range of the antenna element. In [3], seven conductor layers of dipole elements are designed to achieve the required transmission phase range of 360° for a transmitarray antenna. A transmitarray antenna consisting of four identical layers is designed in [4] to increase the transmitarray bandwidth and achieve full transmission phase range of 360° using double square loop element as a unit-cell. In [1], only three identical layers of transmitarray antenna are designed using Jerusalem-cross elements, but limiting the transmission phase range to 335° with 4.4 dB of variation in the transmission magnitude. There are other types of planar lenses used for focusing the electromagnetic waves. Band-pass frequency selective surfaces [5, 6] is one of the most common methods used to design this type of planar lenses.

This chapter investigates the transmission behaviors of M-FSS for transmitarray designs. In contrast to previous publications that studied specific FSS geometries, the goal of this chapter is to reveal the transmission phase limit of M-FSS structures, which will be general for arbitrary FSS geometries. It is shown here that the phase limit of M-FSS is determined by the number of layers, the substrate material, and the separation between layers, and regardless of the element shape. It is revealed that the −1 dB transmission phase limits are 54°, 170°, 308°, and full 360° for single-, double-, triple-, and quad-layer FSS consisting of identical layers, respectively. Furthermore, it is shown that if −3 dB criteria is used, a triple-layer FSS is sufficient to achieve the full 360° phase range.

The validity of the derived phase limits has been verified through numerical simulations of several representative FSS examples. In the full wave simulations, we selected the solver that is most time-efficient for each of the design examples. For single-layer configurations, the MoM solver Ansoft Designer [35] is by far the fastest; therefore, it was used to analyze most of the single-layer configurations. For multi-layer configurations, the 3-D solver CST Microwave Studio software [27] outperforms Ansoft Designer in computation time, thus it is selected for the analysis of all multi-layer structures.

3.1   SINGLE-LAYER FSS ANALYSIS

3.1.1   THEORETICAL ANALYSIS OF SINGLE-LAYER FSS

A single-layer with a conducting element can be considered as a two-port system [1, 2], as shown in Fig. 3.1. It is illuminated from both sides. The incident and reflected plane waves are $\vec{E}_1^+$ and $\vec{E}_1^-$, respectively, at the left-side terminal plane. Similarly, $\vec{E}_2^+$ and $\vec{E}_2^-$ are the incident and reflected plane waves, respectively, at the right-side terminal plane.

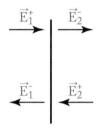

Figure 3.1: Single-layer with a conducting element.

According to the linear two-port networks theory [36], these four complex quantities are related to each other as

$$\begin{bmatrix} E_1^- \\ E_2^- \end{bmatrix} = \begin{bmatrix} S_{11} & S_{12} \\ S_{21} & S_{22} \end{bmatrix} \begin{bmatrix} E_1^+ \\ E_2^+ \end{bmatrix},$$

where $[S]$ is the scattering matrix of the two-port system. Several assumptions and approximations are adopted to derive the useful features of the $[S]$ matrix of the FSS layer.

- Assumption (a): the FSS layer is symmetrical and reciprocal. Then, the following relations are satisfied [36]:

$$S_{11} = S_{22} \quad \text{and} \quad S_{12} = S_{21}. \tag{3.1}$$

- Assumption (b): the FSS layer is lossless. Then, we have [36],

$$|S_{11}|^2 + |S_{21}|^2 = 1, \qquad |S_{12}|^2 + |S_{22}|^2 = 1, \tag{3.2}$$

$$S_{11} S_{12}^* + S_{21} S_{22}^* = 0. \tag{3.3}$$

By substituting Equation (3.1) in Equation (3.3), we get,

$$|S_{11}| e^{j(\angle S_{11})} |S_{21}| e^{-j(\angle S_{21})} + |S_{21}| e^{j(\angle S_{21})} |S_{11}| e^{-j(\angle S_{11})} = 0$$

$$e^{j(\angle S_{11})} e^{-j(\angle S_{21})} + e^{j(\angle S_{21})} e^{-j(\angle S_{11})} = 0$$

$$\angle S_{11} - \angle S_{21} = \pm \frac{\pi}{2}. \tag{3.4}$$

Equation (3.4) shows an interesting observation: the phase difference between the reflected and transmitted waves of any conductor layer is $\pi/2$ regardless of the FSS shape and transmission magnitude.

- Approximation (c): the higher order harmonics of the FSS layer are relatively small and can be neglected. Then, based on the Fresnel reflection and transmission coefficients [37], we get,

$$S_{21} = 1 + S_{11}. \tag{3.5}$$

By substituting Equation (3.4) in Equation (3.5), we get,

$$|S_{21}| e^{j(\angle S_{21})} = 1 + |S_{11}| e^{j(\angle S_{21} \pm \frac{\pi}{2})}$$
$$|S_{21}| - |S_{11}| e^{\pm j \frac{\pi}{2}} = e^{-j(\angle S_{21})}$$
$$|S_{21}| \mp j |S_{11}| = \cos(\angle S_{21}) - j \sin(\angle S_{21}). \tag{3.6}$$

Equation (3.6) can be decomposed into two equations representing the real and imaginary parts, thus,

$$|S_{21}| = \cos(\angle S_{21}), \tag{3.7}$$

and

$$|S_{11}| = \pm \sin(\angle S_{21}). \tag{3.8}$$

It is worthwhile to explain and emphasize Equation (3.7), which reveals the relation between the transmission magnitude and phase. This relation is general and independent from the element shape. It can be demonstrated in a polar diagram, as shown in Fig. 3.2, such that the magnitude represents $|S_{21}|$ and the angle represents $\angle S_{21}$.

The transmission coefficient represents a circle on the polar diagram. The maximum transmission coefficient ($|S_{21}| = 1 = 0$ dB) is achieved only at multiples of $2\pi (\angle S_{21} = 0°, 360°, \ldots)$. In practice, we may accept reduction in the transmission coefficient to a certain limit. Through this chapter, we determine the transmission phase ranges for transmission magnitude limits of -1 dB and -3 dB. Figure 3.2 shows the magnitudes of -1 dB and -3 dB by the dashed green and red circles, respectively. Table 3.1 presents the transmission phase at certain transmission magnitude. Accordingly, the maximum phase range that can be achieved in a single-layer is 54° for -1 dB transmission coefficient and 90° for -3 dB transmission coefficient regardless of the shape of the conducting element.

Equation (3.8) presents the reflection magnitude as a function of the transmission phase. The positive sign is valid when $\sin(\angle S_{21})$ is positive, and vice versa. From Equations (3.1), (3.4), (3.7), and (3.8), the S-parameters of a single-layer FSS can be represented as a function of its

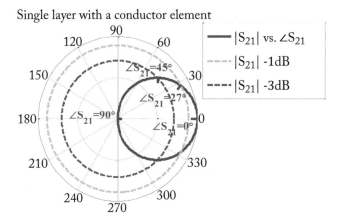

Figure 3.2: Transmission coefficient of a single-layer configuration.

Table 3.1: Transmission phase magnitude relationship of a single conductor layer

$|S_{21}|$ (dB)	0	-1	-3
$\angle S_{21}$ (degrees)	0	±27°	±45°

transmission phase [2]:

$$S_{11} = S_{22} = \sin(\angle S_{21})e^{j\left(\angle S_{21} \pm \frac{\pi}{2}\right)} \tag{3.9}$$

$$S_{12} = S_{21} = \cos(\angle S_{21})e^{j(\angle S_{21})}. \tag{3.10}$$

3.1.2 NUMERICAL DEMONSTRATION OF SINGLE-LAYER FSS

To demonstrate the validity of the assumptions (a) and (b), and the approximation (c), as well as the accuracy of the phase limits in the previous subsection, three representative single-layer unit-cells of a cross dipole, a square loop, and a cross-slot elements are simulated separately at 8.4 GHz.

Figure 3.3 shows the three element unit-cells, with half wavelength periodicity ($P = \lambda_0/2 = 17.86$ mm), variable element length L from 7 mm to 17.5 mm, and element width ($W = 1$ mm). The cross dipole and the square loop elements are simulated using Ansoft Designer software [35]. The cross-slot element is simulated using CST Microwave Studio software [27].

Figure 3.4 shows the transmission magnitudes and phases of the three elements vs. the element length L. Figure 3.5 depicts the transmission magnitudes and phases of the three elements in polar diagrams with the variation of the element length L.

Despite the differences of the transmission coefficient results for the three elements, as shown in Fig. 3.4, the transmission phase magnitude relationship of all three elements agrees

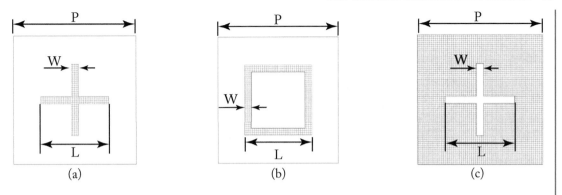

Figure 3.3: Unit-cells of: (a) a cross dipole and (b) a square loop.

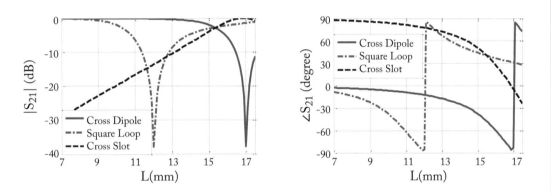

Figure 3.4: Transmission coefficients of the single-layer elements: (a) $|S_{21}|$ and (b) $\angle|S_{21}|$.

with the analytical result of Equation (3.7), as shown in Fig. 3.5. These results demonstrate the generality of the theoretical analysis. Furthermore, it is noticed that the phase range due to the variation of the element length L is not sufficient to cover the complete circle of Fig. 3.2.

It is valuable to realize the differences between the three elements according to Figs. 3.4 and 3.5. Regarding the cross dipole and the square loop elements, when the size is small ($L = 7$ mm), the maximum transmission coefficient is achieved with a phase close to $0°$. This represents a point located on the right edge of the polar diagram with angle equals to $0°$, as shown in Fig. 3.5a,b. By increasing the length L, the transmission magnitude decreases, moving clockwise on the polar diagram, until the element resonates. The resonance considers a full reflection (no transmission) and is represented by a point located at the center of the polar diagram. The cross dipole element resonates when $L \approx \lambda_0/2$, while the square loop element resonates when its perimeter, measured from the center of each side length, is close to one wavelength

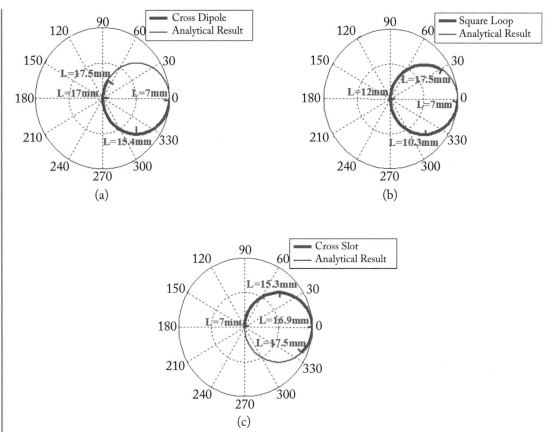

Figure 3.5: Transmission coefficient presented on polar diagrams for: (a) a cross dipole, (b) a square loop, and (c) a cross-slot elements.

$(L \approx \lambda_0/4 + 2W)$. With the continued increase in the length L, the transmission coefficient starts to increase again.

Conversely for the slot-type elements, when the length of the cross-slot element is small $(L = 7$ mm), the minimum transmission coefficient is achieved, which represent the point located at the center of the polar diagram, as shown in Fig. 3.5c. By increasing the length L, the transmission magnitude increases, moving clockwise on the polar diagram, until the element resonates. The resonance for slot-type elements considers a full transmission and is represented by a point located on the right edge of the polar diagram with angle equals to $0°$. The cross-slot element resonates when $L \approx \lambda_0/2$. With the continued increase in the length L, the transmission coefficient starts to decrease.

3.1.3   SINGLE-LAYER OF DOUBLE SQUARE LOOP ELEMENTS

In order to achieve the maximum transmission phase range, we have to select a suitable element shape such that by varying its dimensions within the allowed periodicity of the array unit-cell, the complete circle in Fig. 3.2 is achieved. Selecting an element with double resonance can achieve this demand, because it ensures reaching the center point on the polar diagram twice, or the point on the right edge of the polar diagram for the case of slot-type elements, by varying the element dimensions, which represents a complete circle.

An example of the double resonant elements is the double square loop shape shown in Fig. 3.6 [4]. A single-layer unit-cell of this element is simulated at 8.4 GHz with half wavelength periodicity ($P = \lambda_0/2 = 17.86$ mm) using Ansoft Designer software [35]. The outer loop length L_1 varies from 7 mm to 17.5 mm. The inner loop length L_2 changes with the change of L_1, such that the separation between the two loops is constant ($S = 2.5$ mm). The widths W_1 and W_2 are equal to 0.5 mm.

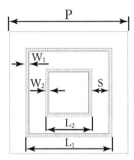

Figure 3.6: Unit-cell of a double square loop element.

Figure 3.7a presents the transmission magnitude and phase vs. the outer side loop length L_1, which shows the two resonant points at $L_1 = 11$ mm and $L_1 = 17.1$ mm. The black solid lines contain the region that achieves transmission coefficient equals to or better than -1 dB, with phase ranges from $-27°$ to $27°$ (phase range of $54°$). Similarly, the red dashed lines contain the region that achieves transmission coefficient equals to or better than -3 dB, with phase ranges from $-45°$ to $45°$ (phase range of $90°$). These results agree with the information of Table 3.1. Figure 3.7b depicts the transmission magnitude and phase in a polar diagram with the variation of the outer side loop length, which conforms to the circle obtained analytically shown in Fig. 3.2. This design is capable of achieving the complete circle in the polar diagram.

It is worthwhile to mention that this analysis also valid for oblique angle of incidence as well, because the derivation in Section 3.1.1 does not use the normal incidence condition. Figure 3.8 depicts the transmission coefficient of the single-layer double square loop element under oblique incidence angle of $30°$ for both the perpendicular and parallel polarizations. Because of oblique incidence and polarization effects, the magnitude and phase curves shift in the rectan-

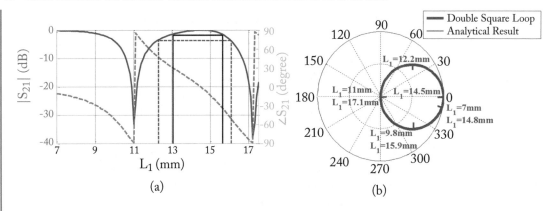

Figure 3.7: Transmission coefficient of the single-layer double square loop element: (a) $|S_{21}|$ and $\angle|S_{21}|$ and (b) polar plot.

gular coordinates (see Fig. 3.8a,b) However, the magnitude and phase relationship remain the same, as shown in the polar diagrams (see Fig. 3.8c, d).

3.1.4   SINGLE CONDUCTOR WITH A SUBSTRATE LAYER

The transmitarray antenna is a multi-layer structure, such that each layer is composed of an array of conductor elements, and these layers are separated by a dielectric substrate or an air gap, or both. The air gap can also be considered as a dielectric substrate with permittivity $\varepsilon_r = 1$.

In order to find the S-Matrix of a multi-layer transmitarray antenna, we have to find first the S-Matrix of one composite layer that is composed of the conductor elements and the dielectric substrate sub-layers, using the S-Matrix of each of these sub-layers.

Figure 3.9 demonstrates a conductor element mounted on a substrate of thickness L_d. This structure can be considered as two sub-layers. The first sub-layer is a conductor and the second sub-layer is a dielectric substrate of thickness L_d. The transmission coefficient of this composite layer can be obtained through the cascading process, as illustrated in detail in the Appendix A. It is worth mentioning that the S-matrix of the dielectric substrate is a function of the dielectric permittivity ε_r and the substrate thickness L_d, while the S-matrix of the conducting element layer is a function of its $\angle S_{21}$ that varies with the variation of the element dimensions.

By varying the transmission phase $\angle S_{21}$ of the conductor layer, we can present, in a polar diagram, the variation of the transmission magnitude and phase of this composite layer. Figure 3.10 shows the polar plot of the transmission coefficient of the single conductor with a substrate layer using numerical analysis for different substrate electrical thickness βL_d with substrate permittivity $\varepsilon_r = 2.5$. It is clear that the transmission phase of the composite layer is a function of the substrate electrical thickness βL_d. The change in the phase appears as a rotation of the transmission coefficient circle of Fig. 3.2 on the polar diagram, such that when the substrate elec-

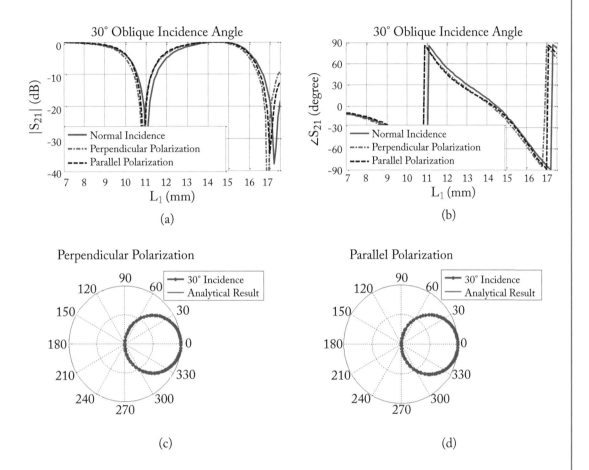

Figure 3.8: Transmission coefficient of the single-layer double square loop element under oblique incidence angle of 30°: (a) $|S_{21}|$, (b) $\angle|S_{21}|$, (c) perpendicular polarization in a polar diagram, and (d) parallel polarization in a polar diagram.

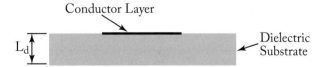

Figure 3.9: Single conductor thick substrate layer.

trical thickness βL_d increases, the circle in the polar diagram moves in the clockwise direction, as shown in Fig. 3.10.

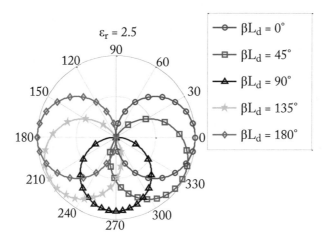

Figure 3.10: Transmission coefficient of a single conductor thick substrate layer for different substrate electrical thickness βL_d at $\varepsilon_r = 2.5$.

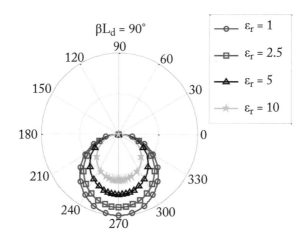

Figure 3.11: Transmission coefficient of a single conductor thick substrate layer for different substrate permitivity ε_r at $\beta L_d = 90°$.

Moreover, the transmission magnitude of the composite layer is also a function of the substrate permittivity, ε_r as a reduction in magnitude at certain electrical thickness values. This reduction appears in the polar diagram as a decrease in the circle diameter, as shown in Fig. 3.10

at $\beta L_d = 90°$. Figure 3.10 shows the change in transmission magnitude (circle diameter) due to the change in the substrate permittivity ε_r at $\beta L_d = 90°$. It is worthwhile to clarify that with the change of permittivity ε_r, a constant electrical thickness βL_d does not means a constant substrate thickness L_d. In order to have a constant electrical thickness βL_d, the substrate thickness L_d should decrease linearly with the increase of $\sqrt{\varepsilon_r}$ because β is also a function of the substrate permittivity $(\beta = 2\pi \sqrt{\varepsilon_r}/\lambda_0)$.

For the special case of substrate permittivity $\varepsilon_r = 1$ which practically represents an air gap, only the transmission phase changes with the change of the air gap L_d, while the transmission magnitude does not degrade at all. This can be understood on the basis that the S-matrix of an air gap has no reflection coefficient (no transmission degradation), as illustrated in the Appendix A. Figure 3.12 shows the change of the transmission phase with the air gap L_d, while the transmission magnitude that is represented by the circle diameter does not change.

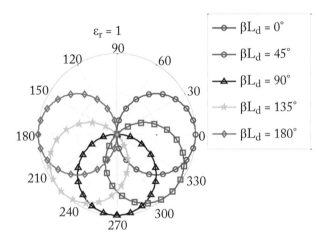

Figure 3.12: Transmission cofficient of a single conductor layer when replacing the thick substrate by an air gap for different βL_d values.

To validate these results, a unit-cell configuration of Fig. 3.9 using the double square loop shape of Fig. 3.6 is simulated at 8.4 GHz with half-wavelength periodicity ($P = \lambda_0/2$) using CST Microwave Studio software [27]. Figure 3.13a shows the transmission coefficient in a polar diagram for different substrate electrical thickness, and Fig. 3.13b shows the polar plot of the transmission coefficient for different substrate permittivity. The two figures confirm the validity of the simulations with the analytical results.

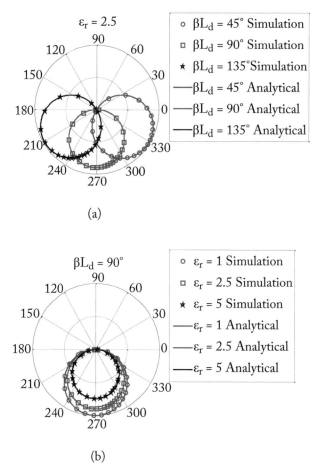

Figure 3.13: Simulation and analytical results of a single conductor thick substrate layer using the double square loop shape: (a) for different substrate electrical thickness βL_d at $\varepsilon_r = 2.5$ and (b) for different substrate permittivity ε_r at $\beta L_d = 90°$.

3.2 DOUBLE-LAYER FSS ANALYSIS

3.2.1 THEORETICAL ANALYSIS OF DOUBLE-LAYER FSS

The double-layer FSS configuration shown in Fig. 3.14 can be considered as three cascaded sections, the top conductor layer, the dielectric substrate, and the bottom conductor layer. We consider both conductor layers identical, so they have the same transmission coefficient phase $\angle S_{21}$. The overall transmission coefficient can be obtained through the cascading process, as illustrated in the Appendix A.

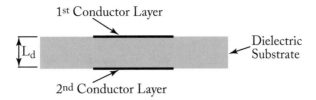

Figure 3.14: Double-layer FFS configuration.

By varying the transmission phase of the conductor layers, we can present in a polar diagram the variation of the transmission magnitude and phase of the double-layer configuration for different substrate permittivity ε_r, as shown in Fig. 3.15, and for different substrate electrical thickness βL_d, as shown in Fig. 3.16. It is observed that the transmission phase magnitude relationships in all cases are symmetric around the vertical axis of the polar diagram.

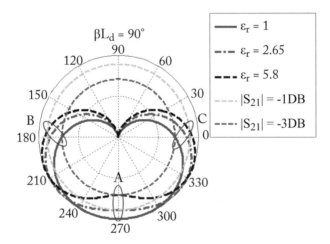

Figure 3.15: Transmission coefficient of the double-layer for different dielectric permitivity by constant electrical thickness of $\beta L_d = 90°$.

From Fig. 3.15, we notice that increasing the substrate permittivity with constant electrical thickness enhances the transmission magnitude at certain phase ranges (around points B and C) but reduces it at another phase range (around point A). The case of electrical thickness of $\beta L_d = 90°$, a substrate permittivity of $\varepsilon_r = 2.65$ has transmission coefficient reduction of -1 dB at point A and the case of a substrate permittivity of $\varepsilon_r = 5.8$ has transmission coefficient reduction of -3 dB at point A.

From Fig. 3.16, the case of substrate electrical thickness of $\beta L_d = 90°$ and permittivity of $\varepsilon_r = 1$ has no transmission coefficient reduction at point A but maximum reduction at points B

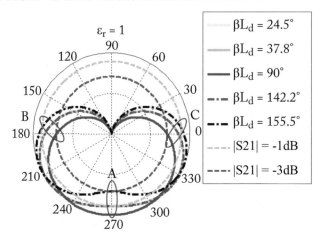

Figure 3.16: Transmission coefficient of the double-layer for different substrate electrical thickness using dielectric permitivity $\varepsilon_r = 1$.

and C (blue solid curve). Increasing or decreasing the electrical thickness away from $\beta L_d = 90°$ enhances the transmission magnitude at points B and C but reduces it at point A. The case of substrate permittivity of $\varepsilon_r = 1$, electrical thickness of $\beta L_d = 37.8°$ and $142.2°$ ($90° \pm 52.2°$) has transmission coefficient reduction of -1 dB at point A, and the case of electrical thickness of $\beta L_d = 24.5°$ and $155.5°$ ($90° \pm 65.5°$) has transmission coefficient reduction of -3 dB at point A.

The changes of the transmission magnitude at points A, B, and C in Fig. 3.15 are equivalent to their changes in Fig. 3.16. Accordingly, we can determine the maximum transmission phase range that can be obtained from any double-layer of conducting elements according to the substrate permittivity and the separation between the conductor layers regardless of the conductor element shape.

Table 3.2 summarizes the transmission phase range for -1 dB and -3 dB transmission coefficient according to the substrate permittivity and electrical thickness. Accordingly, the maximum transmission phase range that can be obtained from a double-layer configuration for -1 dB and -3 dB transmission coefficients are thus $170°$ and $228.5°$, respectively. These phase ranges are still far from the desired phase range of $360°$ in order to support an optimal design of transmitarray antenna.

3.2.2   NUMERICAL DEMONSTRATION OF DOUBLE-LAYER FSS

To demonstrate the accuracy and validity of the above phase limits under the approximation of ignoring the higher-order harmonics, a double-layer unit-cell of the double square loop element of Fig. 3.6 is simulated at 8.4 GHz using CST Microwave Studio software [27].

Table 3.2: Transmission phase range of a double-layer FSS

ε_r	βL_d	Transmission Phase Range (degrees)	
		$|S_{21}| \geq -1\,dB$	$|S_{21}| \geq -3\,dB$
1	90°	128°	180°
	90° ± 52.2°	170°	212.5°
	90° ± 65.5°	–	228.5°
2.65	90°	170°	212.5°
5.8		–	228.5°

Figure 3.17a,b are the simulated transmission magnitudes and phases, respectively, of the double square loop elements vs. the outer square loop side length. Figure 3.17c depicts these results in a polar diagram at different substrate permittivity with constant electrical thickness of $\beta L_d = 90°$. A good agreement between full wave simulation and analytical results can be observed. Furthermore, we notice a small shift from the analytical predictions at some points when the permittivity increases. This is because for constant electrical thickness βL_d, the separation between layers L_d decreases with the increase of the substrate permittivity $\left(L_d = (\beta L_d)\,\lambda_0/2\pi\,\sqrt{\varepsilon_r}\right)$, which hence leads to the increase of the higher-order mode coupling between layers that is ignored in the analytical analysis.

In order to study the reason of the mismatch between the simulation and analytical results and the causes that increase the higher-order mode coupling effect, more simulations have been carried out with different substrate electrical thicknesses and different substrate permittivities.

Figure 3.18 demonstrates the simulation results compared with the analytical results in polar diagrams for different substrate electrical thickness but constant substrate relative permittivity of $\varepsilon_r = 2.5$. We notice the following.

1. When $\beta L_d = 60°$ (see Fig. 3.18a), the mismatch between the simulation and analytical results is high at some points. Besides, the curve based on simulation is not symmetric in the middle of the polar plot like the analytical curve, while the symmetry line is rotated to the right side.

2. When $\beta L_d = 90°$ (see Fig. 3.18b), the mismatch between the simulation and analytical results is smaller compared to the case when $\beta L_d = 60°$. Besides, the curve based on simulation is approximately symmetric in the middle of the polar plot.

3. When $\beta L_d = 120°$ (see Fig. 3.18c), the mismatch between the simulation and analytical results is even smaller compared to the case when $\beta L_d = 90°$. The curve based on simulation is slightly rotated to the left side of the polar plot, this rotation is not high compared to the case when $\beta L_d = 60°$.

4. When $\beta L_d = 270°$ (see Fig. 3.18d), the simulation results well match the analytical results.

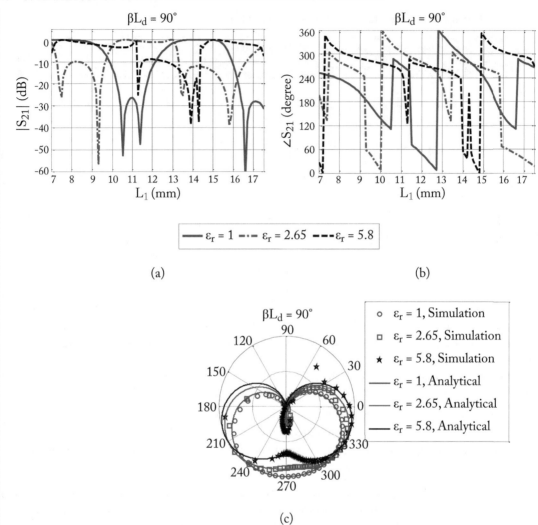

Figure 3.17: Simulation and analytical results of a double-layer configuration using the double square loop element for different dielectric permittivity but constant electrical thickness of $\beta L_d = 90°$: (a) transmission magnitude, (b) transmission phase, and (c) polar plot.

Based on these results, we conclude that the higher-order mode coupling effect decreases with the increase of the separation between layers. To find out whether the substrate permittivity has also some influence on the higher-order mode coupling, two more simulations have been carried out such that the substrate thickness L_d is kept constant while varying the substrate permittivity, as shown in Fig. 3.19.

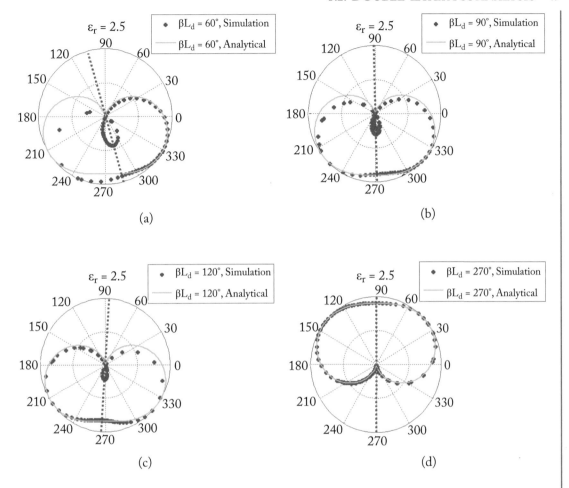

Figure 3.18: Simulation and analytical results of a double-layer configuration using the double square loop element for different electrical thickness but constant substrate permittivity of $\varepsilon_r = 2.5$: (a) $\beta L_d = 60°$, (b) $\beta L_d = 90°$, (c) $\beta L_d = 120°$, and (d) $\beta L_d = 270°$.

Figure 3.19a shows that the coupling effect when $\varepsilon_r = 1$ (air gap separation) is almost negligible. While when the permittivity increases to $\varepsilon_r = 2.5$, there is a shift betw een the simulation and the analytical results. Besides, the curve based on simulation is not symmetric in the middle of the polar plot like the curve based on analytical results, as shown in Fig. 3.19b. These results confirm that the higher-order mode coupling increases when the substrate permittivity increases even for the same substrate thickness.

During the analysis of the single conductor layer in Section 3.1.1, we assumed that the conductor layer is lossless. Additionally, the S-parameters of the dielectric substrate, which are used

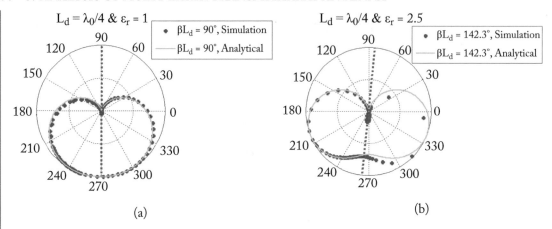

Figure 3.19: Simulation and analytical results of a double-layer configuration using the double square loop element for different substrate permittivity but constant substrate thickness of $L_d = \lambda 0/4$: (a) $\varepsilon_r = 1 \ (\beta L_d = 90°)$ and (b) $\varepsilon_r = 2.5 \ (\beta L_d = 142.3°)$.

in the analytical analysis and are given in the Appendix A, do not consider the substrate losses. Accordingly, it is worthwhile to study the implications when realistic losses are introduced. The conductor and dielectric losses can be considered in the simulation process by selecting practical materials, which include the conductivity of the conductor layers and the loss tangent of the dielectric materials.

Figure 3.20 depicts the simulation results of a double-layer configuration using the double square loop element for both lossless and lossy cases. For the lossless case, PEC is used for the conductor layers and a zero loss tangent substrate of $\varepsilon_r = 2.65$ was used for the dielectric material. While for the lossy case, a copper is used for conductor layers with conductivity of 5.8e7 S/m, and a Taconic substrate of $\varepsilon_r = 2.65$ was used for dielectric materials with loss tangent of 0.0018. It can be noticed that the loss effect is relatively small for practical materials.

3.3   MULTI-LAYER FSS ANALYSIS

3.3.1   ANALYTICAL ANALYSIS OF TRIPLE-LAYER FSS

Since the double-layer FSS cannot achieve the required transmission phase range of 360°, we continue to study the triple-layer FSS. The S-matrix of the triple-layer configuration of Fig. 3.21 can be obtained by cascading two more sections, the dielectric substrate and the third conductor layer, to the double-layer configuration of Fig. 3.14. We consider all conductor layers are identical, so they have the same transmission coefficient phase $\angle S_{21}$.

By varying the transmission phase $\angle S_{21}$ of the three identical conductor layers, we present in a polar diagram the variation of the transmission coefficient of the entire triple-layer configura-

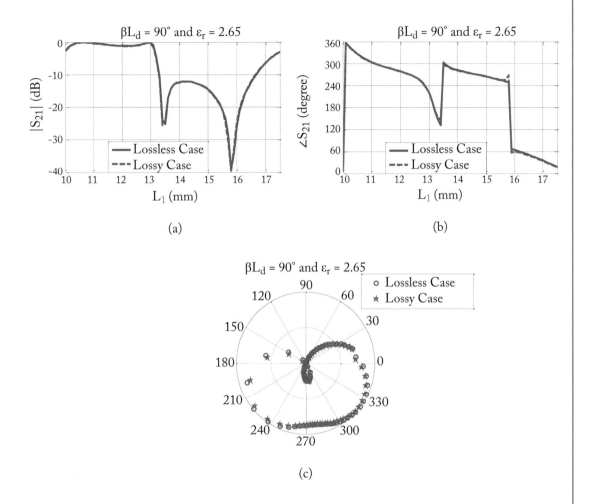

Figure 3.20: Simulation results of a double-layer configuration using the double square loop element for both lossless and lossy materials.

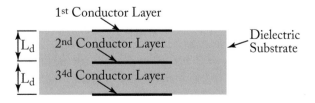

Figure 3.21: Triple-layer FSS configuration.

tion for different substrate electrical thicknesses βL_d with two different substrate permittivities, as shown in Fig. 3.22.

From Fig. 3.22, we notice that at $\beta L_d = 90°$, the transmission phase magnitude relationship in the two polar plots is symmetric around the horizontal axis of the polar diagram. When the electrical thickness βL_d decreases below 90°, the transmission magnitude is reduced at certain phase range around 120°. When the electrical thickness βL_d increases above 90°, the transmission magnitude is reduced at another phase range around 240°. Thus, we conclude that an electrical thickness of $\beta L_d = 90°$ is the optimum value to obtain maximum transmission phase range.

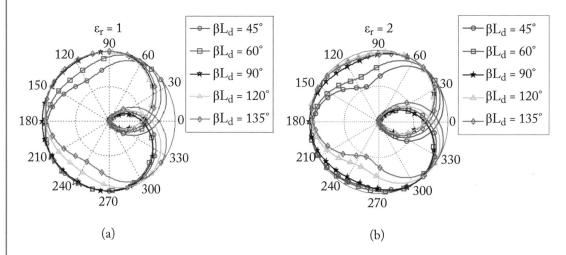

Figure 3.22: Transmission coefficients of the triple layer for different substrate electrical thickness using dielectric permittivity: (a) $\varepsilon_r = 1$ and (b) $\varepsilon_r = 2$.

Figure 3.23 presents in a polar diagram the variation of the transmission coefficient of the entire triple-layer configuration for different substrate permittivities and constant electrical thickness of $\beta L_d = 90°$. From Fig. 3.23, we notice that at $\beta L_d = 90°$, the transmission phase magnitude relationship in all cases are symmetric around the horizontal axis of the polar diagram. The case of $\varepsilon_r = 1$ has the smallest phase range. Increasing the substrate permittivity enhances the transmission magnitude at certain phase ranges (around points C, D, and E) and reduces it at another phase ranges (around points A and B).

A substrate permittivity of $\varepsilon_r = 2$ has transmission coefficient reduction of -1 dB at points A and B. Thus, we consider it the optimal permittivity that achieves maximum phase range for -1 dB transmission coefficient. The case of a substrate permittivity of $\varepsilon_r = 2.5$ has its minimum transmission coefficient of -3 dB at point E. Also, a substrate permittivity of $\varepsilon_r = 4.7$ has its minimum transmission coefficient of -3 dB at points A and B. Consequently, a full phase range of 360° for -3 dB transmission coefficient can be obtained using substrate permittivity be-

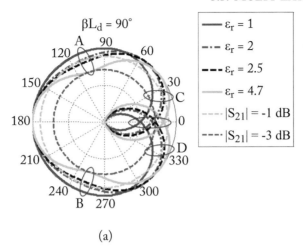

(a)

Figure 3.23: Transmission phase range of a triple-layer FSS with electrical thickness between the conductor layers of $\beta L_d = 90°$.

tween $\varepsilon_r = 2.5$ and $\varepsilon_r = 4.7$. Table 3.3 presents the phase range for -1 dB and -3 dB transmission coefficients with different substrate permittivities. In summary, the maximum transmission phase range that can be obtained using a triple-layer configuration is $308°$ for -1 dB transmission coefficient, and a full transmission phase range of $360°$ for -3 dB transmission coefficient.

Table 3.3: Transmission phase range of a triple-layer FSS with electrical thickness between the conductor layers of $\beta L_d = 90°$

ε_r	Transmission Phase Range (degrees)	
	$|S_{21}| \geq -1\text{dB}$	$|S_{21}| \geq -3\text{dB}$
1	266°	317°
2	308°	352°
2.5	–	360°
4.7	–	360°

3.3.2   NUMERICAL DEMONSTRATION OF TRIPLE-LAYER FSS

A triple-layer FSS consisting of the double square loop element of Fig. 3.6 is simulated at 8.4 GHz using CST Microwave Studio software [27]. Figure 3.24a,b present the simulated transmission magnitudes and phases vs. the outer square loop side length at two different substrate permittivities with constant electrical thickness of $\beta L_d = 90°$.

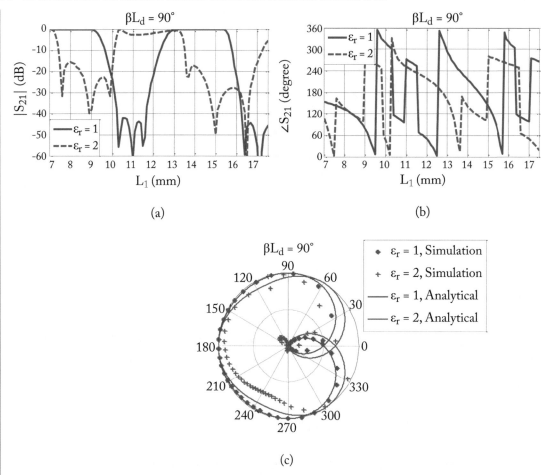

Figure 3.24: Simulation and analytical results of a triple-layer configuration using the double square loop element for different dielectric permittivity but constant electrical thickness of $\beta L_d = 90°$: (a) transmission magnitude, (b) transmission phase, and (c) polar plot.

Figure 3.24c presents the transmission coefficients in a polar diagram. We notice that the numerical results conform well to the analytical results when $\varepsilon_r = 1$. When the substrate permittivity increases, the higher-order mode coupling between layers also increases, leading to a shift between the full wave simulations and the analytical predictions at some points. Nevertheless, the analytical results can provide a good reference for the transmission phase limits.

3.3.3   QUAD-LAYER FSS

For further improvements in the transmission phase range and to avoid the higher-order mode coupling between layers due to the high substrate permittivity, one more conductor layer can be added, as shown in Fig. 3.25.

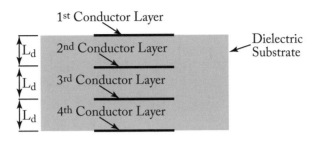

Figure 3.25: Quad-layer FSS configuration.

The polar diagram of Fig. 3.26 illustrates that a full transmission phase range for -1 dB transmission coefficient can be achieved using the quad-layer of conducting elements with substrate permittivity of $\varepsilon_r = 1$ (air gap) and electrical separation between layers of $\beta L_d = 90°$.

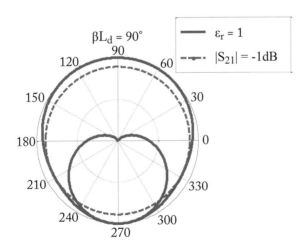

Figure 3.26: Transmission coefficient of the quad-layer FSS for $\beta L_d = 90°$ and $\varepsilon_r = 1$.

A quad-layer FSS consisting of the double square loop element of Fig. 3.6 is simulated at 8.4 GHz using CST Microwave Studio software [27]. Figure 3.27a,b present the simulated transmission magnitude and phase, respectively, vs. the outer square loop side length with $\beta L_d = 90°$ and $\varepsilon_r = 1$. The full wave simulation result in the polar diagram of Fig. 3.27c conforms well to the analytical result.

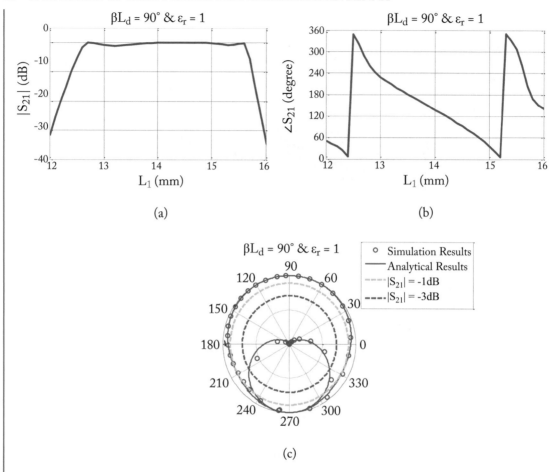

Figure 3.27: Simulation and analytical results of the quad-layer configuration using the double square loop element for $\beta L_d = 90°$ and $\varepsilon_r = 1$: (a) transmission magnitude, (b) transmission phase, and (c) polar plot.

The multi-layer analysis that has been presented in this chapter assumes that the conductor layers are separated by either an air-gap or a dielectric substrate. In the case of air-gap separations, the conducting elements are mostly mounted on a dielectric substrate. The thickness of this dielectric substrate is usually small compared to the air-gap between layers, and its effect can be ignored. Nevertheless, for more accurate analysis the substrate thickness can be considered through rearranging the multi-layer subsections, such that the S-parameters of both the thin dielectric substrate and the air-gap should be considered in the cascading procedure of the multi-layer configuration.

CHAPTER 4

A Quad-layer Transmitarray Antenna Using Slot-type Elements

There are different techniques for transmitarray designs to control the transmission phase of each unit-cell in the array, aiming to obtain a transmission phase range of 360°, while maintaining the high value of the transmission magnitude. These techniques were mentioned in Section 1.3. It is worthwhile to point out that these techniques mostly use printed-type elements mounted on substrate materials [1], [3]–[23]. In this chapter, a cross-slot element is used to design a quad-layer transmitarray antenna. This design has a novelty in using slot-type element with no dielectric substrate, which has two main advantages. The first advantage is its suitability for space applications, because it removes the dielectric substrate that is vulnerary to the extreme temperature change in the outer space. The second advantage is the cost reduction because there is no need to use high performance microwave substrate. A transmitarray antenna has been designed, fabricated, and tested at 11.3 GHz operating frequency. The measured gain of the prototype transmitarray is 23.76 dB. It is observed that the oblique incidence and the wave polarization have strong effect on the transmission coefficient of the slot-type element. Thus, a detailed analysis of the transmitarray considering the oblique incidence angles and the feed polarization conditions is performed. Good agreement between the simulation and measured results is obtained.

4.1 CROSS-SLOT TRANSMITARRAY ANTENNA DESIGN

4.1.1 CROSS-SLOT ELEMENT DESIGN

A unit-cell of a cross-slot element, as shown in Fig. 4.1, is simulated using CST S Microwave Studio software [27] at 11.3 GHz with normal incidence plane wave. Different slot length L_s with periodicity $P = 0.62\lambda_0 = 16.46$ mm and slot width $W = 2$ mm are considered. The purpose of the large unit-cell size of $0.62\lambda_0$ is to gurantee the shortest conductor width between two adjacent slots of not less than 4 mm, which aims to maintain the mechanical strength of the conductor layer. Quad-layer of this cross-slot element is used for this design with separation between layers equals to $H = \lambda_0/4 = 6.64$ mm. This configuration has been selected based on the study of multilayer transmitarrays presented in Chapter 3. This study illustrates that a unit-cell of four identical conductor layers, separated by quarter wavelength air gaps, can achieve a full transmission phase range of 360° for transmission coefficients equal to or better than −1 dB, as shown in Fig. 3.26.

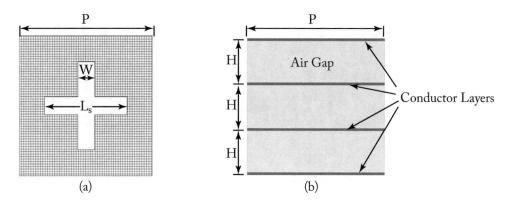

(a)   (b)

Figure 4.1: Cross-slot element unit-cell: (a) top view and (b) side view.

Figure 4.2 shows the transmission magnitude and phase versus the slot length L_s, which confirms the possibility of achieving 360° transmission phase range with transmission coefficient equals to or better than −1 dB.

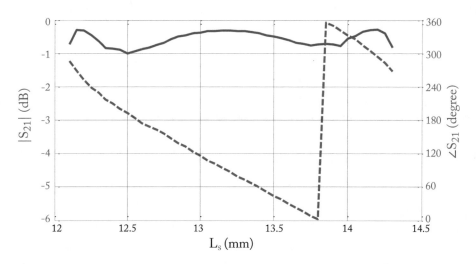

Figure 4.2: Transmission coefficient vs. the slot length L_s for the four identical layers of the unit-cell shown in Fig. 4.1.

4.1.2   TRANSMITARRAY DESIGN AND MEASUREMENTS

The required transmission phase of each transmitarray element is designed to compensate the spatial phase delay from the feed horn to that element, so that a certain phase distribution can be realized to focus the beam at a specific direction. The transmission phase distribution of the ith element was given in Equation (2.1). Once the phase is determined for the ith element, the corresponding slot length can be obtained from Fig. 4.2.

A quad-layer circular aperture transmitarray antenna of diameter $= 13.02\lambda_0 = 34.57$ cm with 325 cross-slot element unit-cell was fabricated for F/D ratio of 0.8. The feed horn is vertically polarized (along y-direction in the xy plane) with q value equals to 6.6. The transmitarray mask is shown in Fig. 4.3a. Figure 4.3b presents a picture of the fabricated quad-layer circular aperture transmitarray.

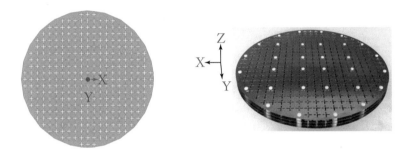

Figure 4.3: A quad-layer circular aperture transmitarray antenna: (a) transmitarray mask and (b) picture of the fabricated quad-layer transmitarray.

The antenna performances of the fabricated prototype were measured using the NSI planar near-field system shown in Fig. 4.4. At 11.3 GHz, the antenna shows a focused beam with a measured gain of 23.76 dB, as shown in Fig. 4.5, leading to an aperture efficiency of 14.2%. The HPBW are 4.6° and 8.8° in the H-plane and E-plane, respectively. The side lobe and cross polarized levels are -13 dB and -30 dB, respectively. Figure 4.6 presents the transmitarray antenna measured gain versus frequency. The maximum measured gain is 24.26 dB and is located at 11.45 GHz. The 1 dB and 3 dB gain bandwidths are 4.2% and 9.4%, respectively.

Using the transmission magnitude and phase properties shown in Fig. 4.2, the array theory [28] is used to calculate the radiation pattern and gain of this transmitarray at 11.3 GHz. These results are compared with the measured data in Fig. 4.5. We noticed approximately 5.2 dB differences in maximum gain and an increase of the side lobes in the measurements compared to the theoretical results. Furthermore, asymmetric beam widths are observed in the measured results, in spite of the symmetry of the elements along the x- and y-directions.

The differences between the measured and theoretical results are caused by the normal incidence approximation usually used for element analysis, which was introduced in Section 2.6.2.

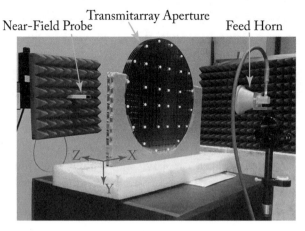

Figure 4.4: Transmitarray antenna setup for a near field measurement.

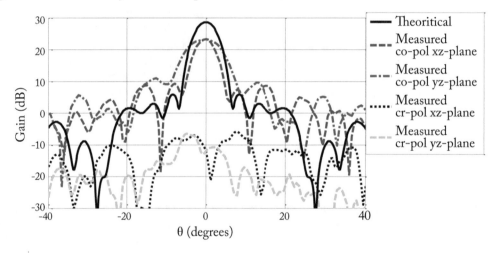

Figure 4.5: Measured and simulated radiation pattern, considering only normal incidence plane wave in the simulation for all array elements.

Therefore, we have carefully studied the effect of the oblique incidence angles and the feed polarization on each array element separately.

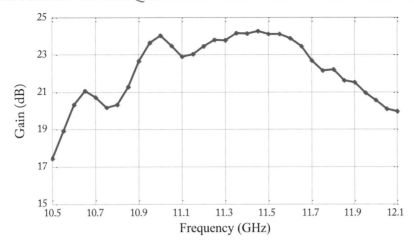

Figure 4.6: Transmitarray antenna measured gain vs. frequency.

4.2   DISCUSSION ON OBLIQUE INCIDENCE AND FEED POLARIZATION EFFECTS

4.2.1   ELEMENT PERFORMANCE UNDER OBLIQUE INCIDENCE

First, we re-simulated the unit-cell element of Fig. 4.1, considering different values of the oblique incidence angle θ, and for the elements along x-axis ($\phi = 0°$) and y-axis ($\phi = 90°$). For y-polarized incidence field, the transmission coefficient of the elements along the x-axis is represented by $T_{\perp\perp}$, and the transmission coefficient of the elements along the y-axis is represented by $T_{\parallel\parallel}$. Here, $T_{\perp\perp}$ and $T_{\parallel\parallel}$ are the perpendicular and parallel transmission coefficient components, respectively, that are obtained from the numerical simulations.

Figure 4.7 shows the variations in the transmission coefficient at different oblique incidence angles, and for y-polarized feed horn. For the elements along the x-axis ($\phi = 0°$), there are almost no variations in the transmission magnitude and phase, except small magnitude reduction at very small slot length L_s. For the elements along the y-axis ($\phi = 90°$), and at high oblique incidence angle ($\theta = 30°$), the transmission coefficient is very poor.

The reason for the transmission reduction is due to the use of the slot-type element shape in a multilayer configuration. Figure 4.8 presents the transmitarray elements illuminated by the feed horn. For vertically polarized feed antenna, the effective part of the cross-slot element is the horizontal slot [38]. Thus, along the x-axis ($\phi = 0°$) and at high oblique angles, each conductor layer hides part of the horizontal slot length of the following layers, which led to small reduction on the transmission coefficient, as shown in Fig. 4.7a. While for elements along the y-axis ($\phi = 90°$), each conductor layer may hide completely the horizontal slot length of the following layers, which led to total reflection.

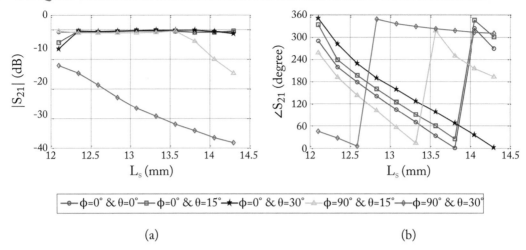

Figure 4.7: Transmission coefficient versus slot length at different oblique incidence angles: (a) transmission magnitude and (b) transmission phase.

As a comparison, the oblique incidence effect is usually smaller when using printed-type elements as in [1, 4]. To demonstrate this point, a quad-layer unit-cell using the printed double square loop element of Fig. 3.6 is simulated using CST Microwave Studio software [27] at both normal incidence and oblique incidence angle of 30°. The unit-cell has periodicity of $0.62\lambda_0$ and total thickness of $0.75\lambda_0$, which is the same as that of the cross-slot element in Fig. 4.1. The dimensions of the double-square loop element are $W_1 = W_2 = 0.5$ mm and $S = 2.5$ mm. The lengths L_1, and L_2 vary such that different transmission phases can be obtained. The transmission coefficient of the double square loop element is shown in Fig. 4.9. The differences between the normal incidence and 30° oblique incidence angle are not significant compared to the case of the cross-slot element shown in Fig. 4.7.

4.2.2   APERTURE DISTRIBUTION AND RADIATION PATTERN

Next, we considered the effect of the oblique incidence angles on all transmitarray elements. Figure 4.10 demonstrates the oblique incidence wave from the feed horn on a sample array element. The feed horn is vertically polarized and is located on the z-axis above the aperture center point by a focal distance F, such that F/D ratio equals to 0.8. Accordingly, each array element is fed by an oblique incidence wave defined by the angles θ and ϕ. The incidence electric field on a certain array element can be defined by the two orthogonal components $E_\perp^i$ and $E_\parallel^i$, as shown in Fig. 4.10. The transmitted electric field components ($E_\perp^t$ and $E_\parallel^t$) from that element can be

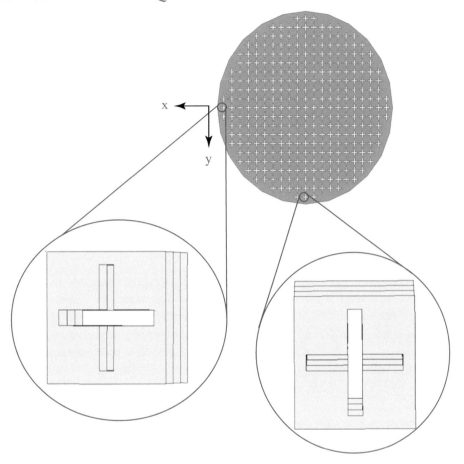

Figure 4.8: The transmitarray elements illumination by the feed horn.

defined as:

$$
\begin{bmatrix} E_\perp^t \\ E_\parallel^t \end{bmatrix} = \begin{bmatrix} T_{\perp\perp} & T_{\perp\parallel} \\ T_{\parallel\perp} & T_{\parallel\parallel} \end{bmatrix} \begin{bmatrix} E_\perp^i \\ E_\parallel^i \end{bmatrix},
\tag{4.1}
$$

where $[T]$ is the transmission coefficient matrix and is obtained from the numerical simulation of the unit-cell element with the consideration of the oblique incidence angles θ and ϕ. The $E_\perp^i$ and $E_\parallel^i$ components are obtained from the equations that describe the radiation pattern of the feed horn as functions of the angles θ and ϕ [29]. The transmitted vertically polarized E_y^t and the transmitted horizontally polarized E_x^t electric field components can then be obtained as follows:

$$
\begin{bmatrix} E_y^t \\ E_x^t \end{bmatrix} = \begin{bmatrix} \cos\phi & \sin\phi\cos\theta \\ -\sin\phi & \cos\phi\cos\theta \end{bmatrix} \begin{bmatrix} E_\perp^t \\ E_\parallel^t \end{bmatrix}.
\tag{4.2}
$$

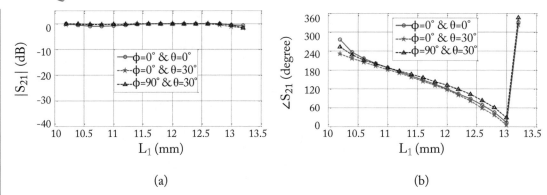

Figure 4.9: Transmission coefficient of the double square loop element at normal incidence and 30° oblique angle: (a) transmission magnitude and (b) transmission phase.

We have re-simulated each element separately at 11.3 GHz taking into account the corresponding oblique incidence angle and the feed polarization conditions. Figure 4.11 compares the magnitudes of the transmitted vertically polarized electric fields with both the normal incidence plane wave approximation and the oblique incidence consideration. For normal incidence approximation, the field distribution is symmetric. While for oblique incidence consideration, we observed a wide field distribution along the x-axis, and narrow field distribution along the y-axis. This explains the wide beam width in the vertical plane cut (yz plane cut) and the reduction of the antenna gain.

We can approximately indicate the minimum incidence angle at which the slots of the following layers are completely hidden. For slot width of $W = 2$ mm and separation between layers $H = 6.64$ mm, the slots of the following layers for the elements along y-axis are completely hidden at oblique angle equal to $\tan^{-1}(2/6.64) = 16.760°$. This occurs for the closest four elements to the edge along y-axis, as shown in Fig. 4.11b.

The radiation pattern and gain have been re-calculated using the array theory with oblique incidence excitation [28]. The results are depicted in Fig. 4.12, which shows much better agreement with the measurements. The beam widths of the measured vertical and horizontal plane cuts match the theoretical calculations very well. The side lobe level of the measured horizontal plane cut conforms to the theoretical results. However, the first side lobe level of the calculated vertical plane cut is higher than that of the measured results. This difference makes the theoretical gain a little less than the measured gain by 0.55 dB at 11.3 GHz. The theoretical and measured gains are 23.21 dB and 23.76 dB at 11.3 GHz, respectively. Table 4.1 presents a comparison of transmitarray measured and simulated performance.

We consider the discrepancy between the theoretical and measured results is due to the periodic boundary condition approximations (infinite array approximations) during the simulation process of each element, which was introduced in Section 2.6.2. This approximation assumes that

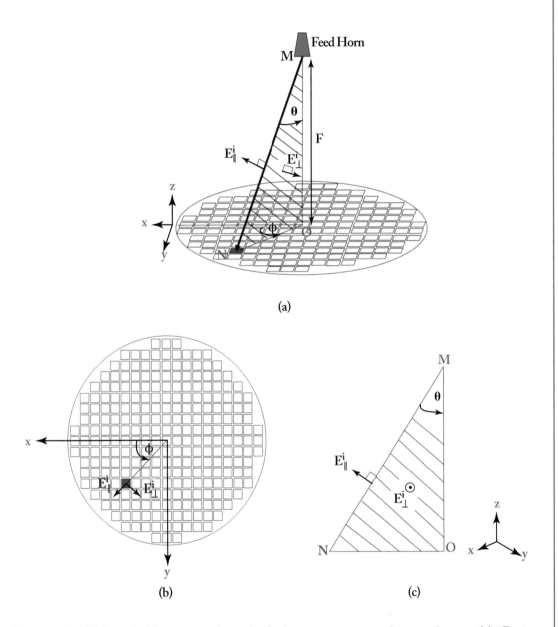

Figure 4.10: Oblique incidence wave from the feed antenna on a sample array element: (a) 3D view, (b) top view (x-y plane), and (c) plane of incidence view.

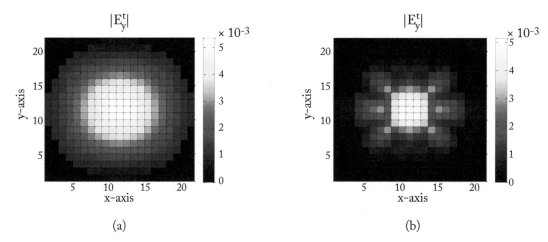

(a)   (b)

Figure 4.11: Transmitted vertically polarized electric field: (a) with normal incidence plane wave approximation and (b) with oblique incidence.

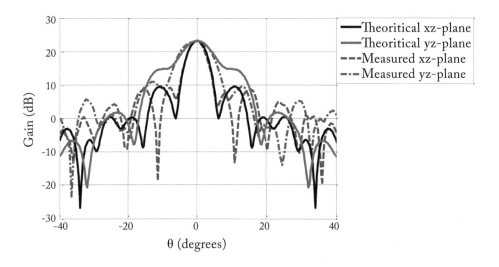

Figure 4.12: Measured and simulated radiation pattern, considering both the oblique incidence angles and the feed polarization in the simulation of each element of the array.

an infinite array of elements have the same oblique incidence and polarization. This disagreement is small for the horizontal plane cut (xz plane, $\phi = 0°$), because the sensitivity of the cross-slot element to the oblique incidence along perpendicular polarization is small as shown in Fig. 4.7. While due to the slot-type element configuration, the element is very sensitive to oblique incidence along parallel polarization, as shown in Fig. 4.7, and hence it is sensitive to the periodic boundary condition approximations in the simulation processes.

Table 4.1: Comparison of transmitarray measured and simulated performance

	HPBW (H-Plane)	HPBW (E-Plane)	Gain
Normal incidence	4.6°	4.6°	28.56. dB
Oblique incidence	4.6°	8.8°	23.21 dB
Measured	4.6°	8.8°	23.76 dB

CHAPTER 5

Design of Triple-layer Transmitarray Antennas

The transmission phase magnitude relationship of multi-layer transmitarrays has been studied in Chapter 3. This study confirms that a multi-layer configuration is required to design a transmitarray. A seven conductor layers transmitarray antenna is presented in [3] using dipole elements to achieve the full transmission phase range of 360°. In [4], a four identical layer transmitarray using double square loop element achieves the full transmission phase range of 360°. To further reduce the number of conductor layers, a three-layer transmitarray antenna is designed using Jerusalem-cross elements in [1], but limiting the transmission phase range to 335° with 4.4 dB of variation in the transmission magnitude. A reconfigurable triple-layer transmitarray achieves 360° phase range using varactor diodes in [12]. There are some triple-layer designs where discrete phases states (0°/180° for 1-bit, and 0°/90°/180°/270° for 2-bit) are used for beam steering application, at the expense of reducing the overall antenna gain [10, 13, 17].

Hence, in order to reduce the antenna cost and complexity, a challenge is to achieve a full transmission phase range of 360° using fewer conductor layers while avoiding the reduction of the element transmission magnitude and maintaining the overall performance of the transmitarray antenna. This chapter presents three different methods to design triple-layer transmitarray antennas.

The first method aims to reduce the contribution of the elements, which have low transmission magnitudes, on the overall antenna loss. The transmission phase of the center element of the transmitarray aperture is optimized, such that the transmission magnitudes of this element and the elements closer to the aperture center equal 1 (0 dB), while keeping the elements with smaller transmission coefficients away as much as possible from the aperture center. This is because the radiation pattern of the feed antenna is directed with its maximum power to the center of the transmitarray aperture, while the feed illumination decreases away from the aperture center. Based on this method a prototype of high gain transmitarray antenna is designed, fabricated, and tested.

The second method is based on the use of transmitarray elements with non-identical layers to cover the transmission phase region that has degradation in the transmission magnitude when using identical layers. For the unit-cells with non-identical layers, either the element has different conductor layer shape, or they have the same shape but different dimensions. This method has been analytically studied and verified through numerical simulations. Moreover, the sensitivity of the unit-cell to manufacturing accuracy has been clarified, and different ways to minimize this impact have been discussed.

The third method relies on using two groups of double-layer unit-cells, which have different thicknesses. Each unit-cell group has a transmission phase range differ from that of the other group. A full transmission phase range of 360° is obtained when combining the transmission phase ranges of the two groups of unit-cells. Each unit-cell group has been analytically studied and verified through numerical simulations. However, there is a challenge for accurate use of periodic boundary conditions in the numerical simulations of combined two different unit-cells.

5.1   IDENTICAL TRIPLE-LAYER TRANSMITARRAY ANTENNA

In this section, we present a novel design of a triple-layer transmitarray antenna. Using spiral dipole elements, a full phase range of 360° is achieved for a transmission magnitude equal to or better than −4.2 dB. Furthermore, the element phase at the center of the transmitarray aperture is selected deliberately, in order to reduce the effects of the lossy elements with low transmission coefficient magnitudes on the antenna gain, leading to an average element loss as low as 0.49 dB.

This triple-layer transmitarray antenna using spiral-dipole elements has been fabricated and tested for X-band operation. The measured gain of the transmitarray prototype is 28.9 dB at 11.3 GHz, and the aperture efficiency (ϵ_{ap}) is 30%. The measured 1 dB and 3 dB gain bandwidths are 9% and 19.4%, respectively, which are considered broadband performances as compared with the published designs in [4, 16–19].

5.1.1   SPIRAL DIPOLE ELEMENT DESIGN

Based on the analytical analysis that presented in Chapter 3, varying the length of a conventional dipole element within a limited unit-cell size (such as $\lambda_0/2$) is not sufficient to cover a full transmission phase range [2]. An extension of the dipole length can be done by bending the dipole arm. Moreover, to maintain the symmetry along the x- and y-axes, a spiral dipole shape is designed. Figure 5.1 shows the conventional cross-dipole and spiral-dipole elements in three identical layer configurations, with unit-cell periodicity of $P = 0.6\lambda_0 = 15.93$ mm. The free space wavelength λ_0 is computed at 11.3 GHz. The element of each layer is mounted on a thin dielectric substrate of thickness $T = 0.5$ mm and permittivity $\varepsilon_r = 2.574$. The separation between layers is equal to $H = 6$ mm, such that the total separation between two layers is close to a quarter wavelength ($H + T \approx \lambda_0/4 = 6.64$ mm).

The two elements are simulated using CST Microwave Studio software [27] at 11.3 GHz with normal incidence plane wave. Various dimensions of length L and width $W = 0.1L$ are considered. The transmission magnitudes and phases of the two elements vs. the element dimension L are presented in Fig. 5.2a,b. It is worthwhile to emphasize the relation between the transmission

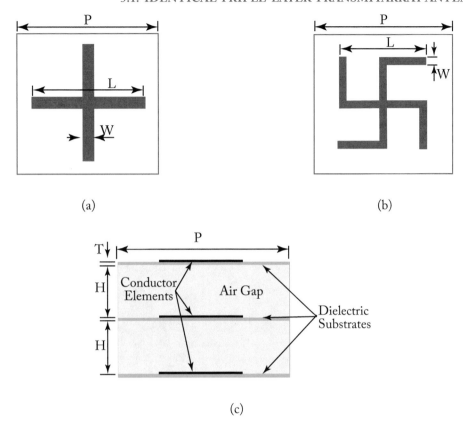

Figure 5.1: Triple-layer unit-cell: (a) cross-dipole shape, (b) spiral-dipole shape, and (c) triple-layer configuration of the unit-cell.

magnitude and phase, as shown in a polar diagram in Fig. 5.2c,d, where the magnitude represents $|S_{21}|$ and the angle represents $\angle S_{21}$.

The phase range of varying the cross-dipole element length from $L = 6$ mm to $L = 15.9$ mm, but neglecting the region where the element resonates (from $L = 9.6$ mm to $L = 12.3$ mm), is not sufficient to cover the full range of 360°. Since the length of the cross dipole is doubled when using the spiral dipole, the transmission magnitude and phase performances of the cross dipole at certain length L is achieved at half dimension $L/2$ when using the spiral dipole. Therefore, the right side curves of the cross dipole magnitude and phase shown in Fig. 5.2a,b are shifted to smaller dimensions when using the spiral dipole. This allows a wider range of varying the element dimension L in the unit-cell to achieve the required phase range of 360°.

A full transmission phase range of 360° is achieved with the use of the spiral dipole element with transmission magnitude equal to or better than −4.2 dB, as shown in Fig. 5.2. Furthermore, a

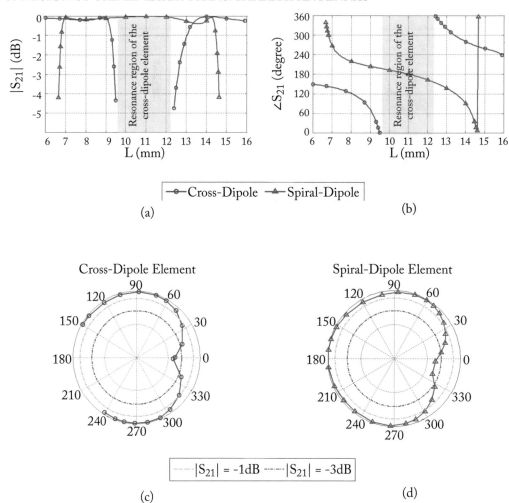

Figure 5.2: Transmission coefficients of the triple-layer cross-dipole and spiral-dipole elements: (a) magnitude vs. the element dimension L, (b) phase vs. the element dimension L, (c) phase magnitude relation of the cross-slot element in a polar diagram, and (d) phase magnitude relation of the cross-slot element in a polar diagram.

$270°$ phase range is achieved with magnitude better than -1 dB and a $320°$ phase range is achieved with magnitude better than -3 dB. The element width of $W = 0.1L$ is selected to maintain a large range of variation in the element length L, and hence a more linear slope is achieved, as shown in Fig. 5.2b. The length L varies between 6.65 mm and 14.65 mm to obtain the full transmission phase range of $360°$, which makes the design less sensitive to manufacturing error.

It is usually assumed in the design of transmitarray antennas that the feed signal is normally incident on all elements, although the majority of the elements are actually illuminated under oblique incidence angles. Thus, it is worthy to evaluate the behavior of the spiral dipole element under oblique incidence. Figure 5.3 depicts the variations in the transmission magnitude and phase at different oblique incidence angles and for y-polarized incidence wave. The parameters ϕ and θ are the azimuth and elevation angles of the incidence wave, respectively. For the E-plane ($\phi = 90°$), there are almost no variations in the transmission magnitude and phase, except small changes at large element dimensions ($L > 13$ mm) and with oblique incidence as high as $\theta = 30°$. For the H-plane ($\phi = 0°$), and with the increase of the oblique incidence angle θ, we noticed not only phase changes but also magnitude reduces at certain values of the element dimension L.

5.1.2   TRANSMITARRAY DESIGN

The phase distribution of the transmitarray aperture was discussed in Section 2.1, and the transmission phase ψ_i for the ith element was given in Equation (2.1) as:

$$\psi_i = k \left(R_i - \vec{r}_i \cdot \hat{r}_o \right) + \psi_0, \tag{5.1}$$

where k is the propagation constant, R_i is the distance from the feed horn to the ith element, $\vec{r}_i$ is the position vector of the ith element, and $\hat{r}_o$ is the main beam unit vector, as shown in Fig. 5.4. For a transmitarray with a main beam at the broadside direction, $\vec{r}_i \cdot \hat{r}_o = 0$. The phase constant ψ_0 is selected to drive the reference phase at the aperture center ψ_c to a certain value. Once the ith element phase is determined, the corresponding spiral-dipole element dimension L can be obtained from Fig. 5.2b. Equation (5.1) doesn't consider the phase of the horn pattern, the oblique incidence and the feed polarization, while an approximation is usually used in practical designs in determining the phase of the unit-cell element for simplicity.

A triple-layer circular aperture transmitarray antenna of diameter $= 16.2\lambda_0 = 43.01$ cm using the spiral-dipole elements was designed for an F/D ratio of 0.8. It includes 537 elements. The feed horn is vertically polarized (along the y-direction in the xy plane) with a gain equal to 15.9 dB at 11.3 GHz. The feed horn pattern is approximately modeled as, $\cos^q(\theta)$, where $q = 6.6$. Referring to Fig. 5.2d, the transmission phase of the transmitarray center element is selected at $\psi_c = 55°$, which has a transmission magnitude equal to 1 (0 dB). Consequently, when the element is away from the aperture center, it needs a transmission phase larger than $55°$, thus moving counterclockwise in the polar diagram of Fig. 5.2d. The transmission magnitudes of these elements are still closer to 1 (0 dB) until the element transmission phase is larger than $300°$, which has a transmission magnitude less than -1 dB (see Fig. 5.2d). The aim of this phase distribution is to keep the lossy elements, which have low transmission magnitudes, away as much as possible from the aperture center, thus reducing their contribution to the average element loss. This is because the radiation pattern of the feed antenna is directed with its maximum power to the center of the transmitarray aperture, while the feed illumination decreases away from the aperture center.

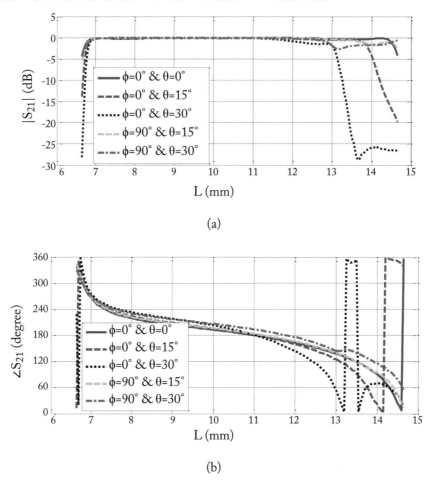

Figure 5.3: Transmission coefficient vs. element dimension L under different incident angles: (a) magnitudes and (b) phases.

The average element loss is calculated for different values of the center element phase ψ_c, as shown in Fig. 5.5. The average element loss calculation was given in Equation (2.42). It is observed from Fig. 5.5 that the average element loss is minimum with 0.49 dB when $\psi_c = 55°$, as expected from previous discussion.

Figure 5.6a presents a top view picture of the fabricated transmitarray aperture. The transmission phase distribution of the transmitarray is shown in Fig. 5.6b with element phase of 55° at the aperture center. Figure 5.6c presents the transmission magnitude distribution, which demonstrates that the elements with high transmission magnitudes are close to the aperture center. Figure 5.6d presents the relative illumination from the feed on the transmitarray elements. It il-

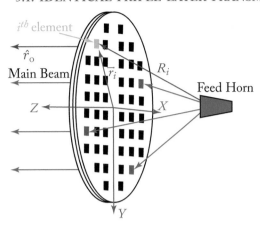

Figure 5.4: Typical geometry of a printed transmitarray antenna.

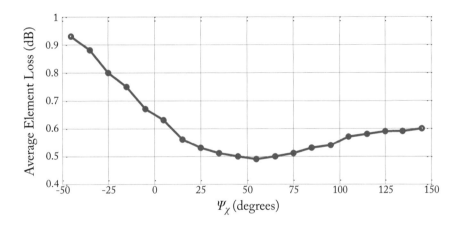

Figure 5.5: Transmitarray average element loss vs. aperture center element phase.

lustrates the concentration of the feed illumination around the aperture center, with illumination taper at the edge of the transmitarray aperture equal to -10.2 dB.

5.1.3 EXPERIMENT AND DISCUSSION

The NSI planar near-field system is used to measure the antenna performances of the fabricated prototype, as shown in Fig. 5.7. At 11.3 GHz, the antenna achieves a focused beam, as shown in Fig. 5.8. The measured directivity and gain at 11.3 GHz are 30.2 dB and 28.9 dB, respectively. Thus, the measured radiation efficiency (gain over directivity ratio) is equal to 74%. The HPBW are 4.0° and 5.0° in the E-plane (yz plane) and H-plane (xz plane), respectively. The side lobe

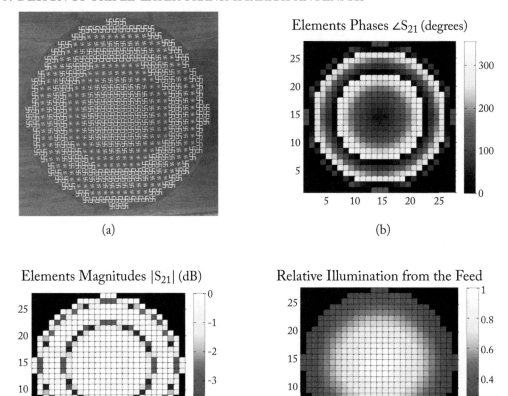

Figure 5.6: Circular aperture transmitarray antenna: (a) top view picture of the transmitarray aperture, (b) elements transmission phase distribution, (c) elements transmission magnitude distribution, and (d) relative illumination from the feed on the transmitarray elements.

and cross polarized levels are −21 dB and −27 dB, respectively, in both planes. The corresponding aperture efficiency ϵ_{ap} is calculated using:

$$\epsilon_{ap} = \frac{G}{D_{\max}}, \qquad D_{\max} = \frac{4\pi A}{\lambda_0^2}, \tag{5.2}$$

where G is the measured gain, $D_{\max}$ is the maximum directivity, A is the area of the antenna aperture, and λ_0 is the free space wavelength. The aperture efficiency is found to be 30%. The

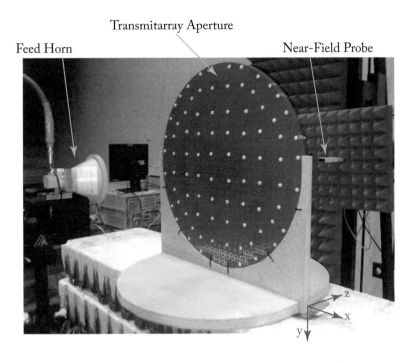

Figure 5.7: Transmitarray antenna setup for a near field measurement.

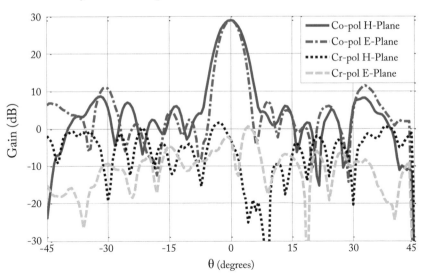

Figure 5.8: Measured H-plane (xz plane) and E-plane (yz plane) radiation patterns.

pattern beam width in E-plane and H-plane are different due to the oblique incidence angle and the linear polarization of the feed horn.

The measured gain of the transmitarray antenna vs. frequency is presented in Fig. 5.9. The 1 dB and 3 dB gain bandwidths are 9% and 19.4%, respectively, which are considered broadband performances achieved using a triple-layer configuration. For all frequencies within the 3 dB gain bandwidth, the radiation patterns have side lobe and cross polarization levels better than −14 dB and −21 dB, respectively. Table 5.1 compares these results with recent published work.

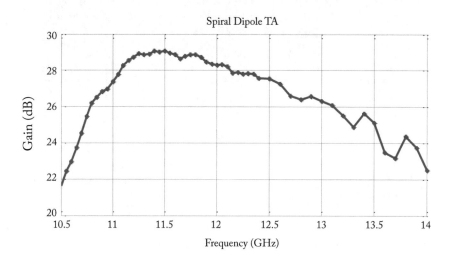

Figure 5.9: Transmitarray antenna gain vs. frequency.

Table 5.1: Comparison of current results with recent published work

Reference #	Frequency (GHz)	# of Layers	Gain (dB)	ϵ_{ap} (%)	1dB Gain BW (%)	3dB Gain BW (%)
This work	11.3	3	28.9	30	9	19.4
[4]*	30.25	4	28.59	35.6	7.5	–
[19]*	9.8	4	22.7	15.4	–	15.8
[17]	60	3	23.90	17	7.1	–
			22.30	12.9	7.6	–
[18]**	5	7	–	–	–	10

* The aperture efficiency is calculated based on the maximum gain and the aperture dimensions available.
** An element design.

The advantages of the transmitarray antenna using spiral-dipole elements can be summarized as follows.

(a) The spiral-dipole element achieves a 360° transmission phase range with a transmission magnitude better than or equal to −4.2 dB. In addition, the reference phase is optimized to reduce the average element loss to be as low as 0.49 dB, which is a good result for a triple-layer configuration.

(b) The spiral-dipole element has large range of variation in the element dimensions, creating a more linear slope for the phase, which makes the design less sensitive to manufacturing error.

(c) Broadbands of 9% at 1 dB-gain and 19.4% at 3 dB-gain are achieved using the spiral-dipole elements in a triple-layer transmitarray design.

5.2   NON-IDENTICAL TRIPLE-LAYER TRANSMITARRAY ANTENNA

The multi-layer analysis, which has been presented in Chapter 3, assumes that all conductor layers are identical. Accordingly, we considered the element shape and dimensions are the same in each layer, and thus the individual transmission phase of each layer $\angle S_{21}$ in the unit-cell equals to those of the other layers. As a result of this study, it was concluded that the quad-layer transmitarray is the best choice to obtain a full transmission phase range of 360°.

This section studies the multi-layer configuration in more details through the analysis for the case of non-identical layers. It also aims to determine the possibility of obtaining a full transmission phase range of 360° with less number of layers using non-identical layer configuration. For the unit-cells with non-identical layers, either the element has different conductor layer shape, or they have the same shape but different dimensions. Hence, each conductor layer in the unit-cell has its own transmission phase value $\angle S_{21}$ that may differ from those of the other layers.

5.2.1   NON-IDENTICAL DOUBLE-LAYER FSS ANALYSIS

The S-matrix of a non-identical double-layer FSS configuration shown in Fig. 5.10 has two degrees of freedom with respect to the element change. At constant substrate parameters (permittivity and thickness), the element dimensions of each conductor layer, and hence the conductor transmission phase $\angle S_{21}$, changes independently from that of the other conductor layer.

By varying the transmission phases of the two conductor layers independently, we can present in a polar diagram all possible variations in the transmission magnitude and phase of the non-identical double-layer configuration, as shown in Fig. 5.11. Two cases of different substrate permittivities, when $\varepsilon_r = 1$ (air gap) and $\varepsilon_r = 2.5$, are presented in Figs. 5.11a,b, respectively. The blue area represents all possible transmission coefficient values that can be obtained when the transmission phase of the two conductor layers vary independently (non-identical case). The red

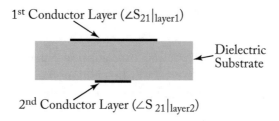

Figure 5.10: Non-identical double-layer FSS configuration.

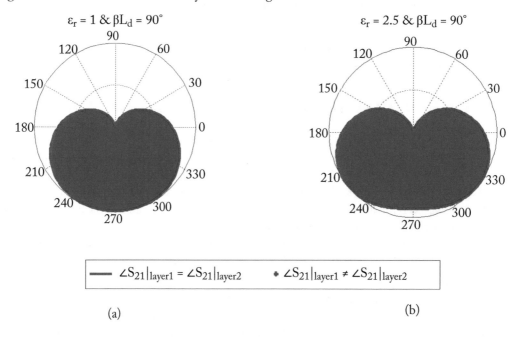

(a)   (b)

Figure 5.11: Non-identical double-layer FSS configuration: (a) $\varepsilon_r = 1$ and (b) $\varepsilon_r = 2.5$.

curve represents the case when the two conductor layers are identical. It is worthwhile to notice that the maximum transmission phase range of a double-layer configuration is obtained when the two conductor layers are identical.

5.2.2 NON-IDENTICAL TRIPLE-LAYER FSS ANALYSIS

For a non-identical triple-layer FSS, the S-matrix could have three degree of freedoms with respect to the element change. The element dimensions of each conductor layer, and hence the conductor transmission phase $\angle S_{21}$, could change independently from those of the other conductor layers. However, for simplicity and to maintain symmetry in the design, we keep the first and third layers identical, such that $(\angle S_{21}|_{layer1} = \angle S_{21}|_{layer3} \neq \angle S_{21}|_{layer2})$, as shown in Fig. 5.12. Accordingly, this configuration still has two degrees of freedom with respect to the element change.

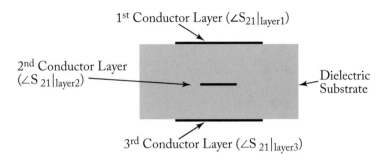

Figure 5.12: Non-identical triple-layer FSS configuration.

By varying the transmission phases of the three conductor layers $(\angle S_{21}|_{layer1} = \angle S_{21}|_{layer3} \neq \angle S_{21}|_{layer2})$, Fig. 5.13 presents all possible values in the transmission magnitude and phase for the triple-layer configuration. Two cases of different substrate permittivity, when $\varepsilon_r = 1$ (air gap) and $\varepsilon_r = 2.5$, are presented in Figs. 5.13a,b, respectively. The blue area represents all possible transmission coefficient values that can be obtained when the transmission phase of the second conductor layer varies independently from the other two conductor layers (non-identical case). The red curve represents the case when the three conductor layers are identical.

The results of Fig. 5.13 reveal the imp ortance of the non-identical triple-layer configuration, which clarify that the transmission phase magnitude relationship can be controlled and a full transmission phase range of 360° can be obtained with high transmission magnitude values. From Fig. 5.13a, we can specify an ideal target using the non-identical triple-layer configuration, which is shown in Fig. 5.14. A full transmission phase range of 360° can be obtained using air-gap separation between layers ($\varepsilon_r = 1$) with electrical thickness of $\beta L_d = 90°$ as follows.

(a) A transmission phase range from 70° to 290° can be obtained using identical conductor layers, $(\angle S_{21}|_{layer1} = \angle S_{21}|_{layer3} = \angle S_{21}|_{layer2})$, because the transmission magnitude of the identical triple-layer configuration is very close to 0 dB (better than or equal to −0.15 dB) in that phase range.

(b) A transmission phase range from −70° to 70° can be obtained using non-identical conductor layers, $(\angle S_{21}|_{layer1} = \angle S_{21}|_{layer3} \neq \angle S_{21}|_{layer2})$, such that each conductor layer is individually designed, while keeping the first and third layers identical, and maintaining the transmission magnitude close to 1 (0 dB), as shown in Fig. 5.14.

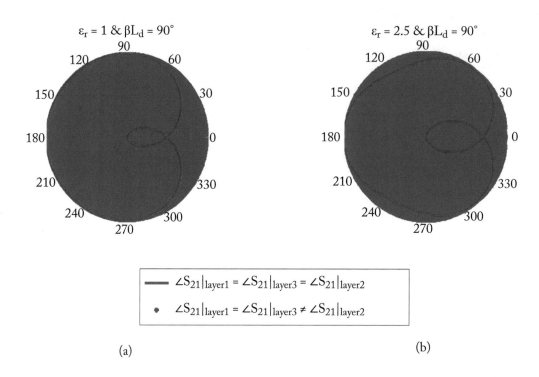

Figure 5.13: Non-identical triple-layer FSS configuration: (a) $\varepsilon_r = 1$ and (b) $\varepsilon_r = 2.5$.

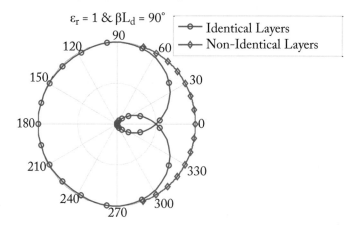

Figure 5.14: Ideal full transmission phase range of 360° of a triple-layer FSS configuration using a combination of both identical and non-identical layers with air-gap separation between layers ($\varepsilon_r = 1$) and electrical thickness of $\beta L_d = 90°$.

In order to clarify the possibility of implementing the ideal target in Fig. 5.14, we have carefully studied the impacts of both the quantization phase error and the manufacturing tolerance of the non-identical triple-layer configuration on the transmission coefficient of the unit-cell.

Quantization Phase Effect and Numerical Demonstration

As we mentioned in Section 2.6.1, the change in element dimensions depends on the manufacturing precision, and hence, a continuous phase control is not possible. Accordingly, it is worth studying the non-identical triple-layer configuration of Fig. 5.12, taking into account the phase quantization of each conductor layer. Although the relation between the transmission phase of each layer vs. the element change is not linear, for simplicity we assume constant quantization phase values. Moreover, the transmission coefficient of a single conductor layer is located only in the first and fourth quadrants of the polar diagram, as shown in Fig. 3.2 and based on Equation (3.7). Thus, the transmission phase of a single-layer varies only between $-90°$ and $90°$.

Figure 5.15 demonstrates the transmission coefficient of a non-identical triple-layer configuration with four different quantization phase values. The conductor layers are separated by an air gap ($\varepsilon_r = 1$) with electrical thickness equal to $\beta L_d = 90°$. We can observe that with the increase in the quantization phase of each layer, the density of the transmission coefficient points in the polar diagram decreases. However, our concern is in the transmission phase that ranges from $-70°$ to $70°$ with transmission magnitude close to 1 (0 dB). The transmission coefficient values in that phase range are still acceptable with quantization phase that may reach $5°$ in each layer, as shown in Fig. 5.15d.

To demonstrate the validity of this analysis and these results, a triple-layer unit-cell of the double square loop shape of Fig. 3.6 is simulated at 11.3 GHz with periodicity ($P = 0.6\lambda_0 = 15.93$ mm) using CST Microwave Studio software [27]. The outer loop length L_1 of each layer varies from $L_1 = 9$ mm to $L_1 = 15.9$ mm with unit step of 0.1 mm, and the inner loop length L_2 changes with the change of L_1. The element of the second layer changes independently of the other elements of the first and third layers, while maintaining these elements of the first and third layers identical.

Figure 5.16 demonstrates the simulation results of all transmission coefficient points. The point distribution and density are different from Fig. 5.15 because of the nonlinear relation between the element dimension and the element phase. Nevertheless, it confirms the possibility of obtaining transmission magnitude values close to 1 (0 dB) with transmission phase that ranges from $-70°$ to $70°$ using the non-identical triple-layer configuration of Fig. 5.12. It is important to mention that a manufacturing accuracy of 0.1 mm, which is used in this simulation, is considered a practical value using the printed circuit board technology.

Phase Sensitivity Analysis

As mentioned in Section 2.6.3, errors in transmission phase of the unit-cell element may arise due to manufacturing tolerances in the etching process of the array elements, and the effect of

$\angle S_{21}|_{layer1} = \angle S_{21}|_{layer3} = \angle S_{21}|_{layer2}$

$\angle S_{21}|_{layer1} = \angle S_{21}|_{layer3} \neq \angle S_{21}|_{layer2}$

Quantization Phase of Each Layer = 0.5°

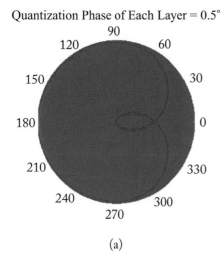

(a)

Quantization Phase of Each Layer = 2°

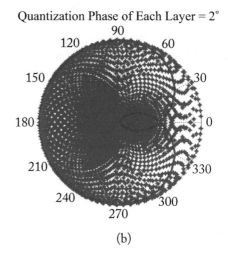

(b)

Quantization Phase of Each Layer = 3°

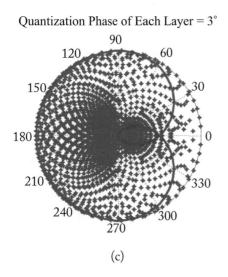

(c)

Quantization Phase of Each Layer = 5°

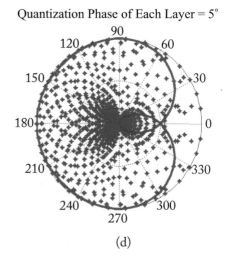

(d)

Figure 5.15: Transmission coefficient of a non-identical, triple-layer FSS configuration with $\varepsilon_r = 1$ and $\beta L_d = 90°$ for quantization phase of each conductor layer equal to: (a) 0.5°, (b) 2°, (c) 3°, and (d) 5°.

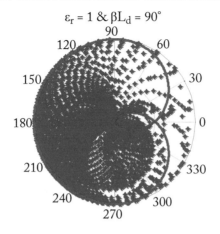

Figure 5.16: Simulation results of a non-identical, triple-layer FSS configuration with $\varepsilon_r = 1$ and $\beta L_d = 90°$ using the double square loop element at 11.3 GHz.

this random phase error on the antenna gain was presented in Section 2.6.3. However, in transmitarray antennas, the manufacturing tolerances in the etching process may affect not only in the transmission phase but also in transmission magnitude, which in turn increases the average element loss.

For the case of identical multi-layer unit-cell, errors in transmission phase and magnitude due to manufacturing tolerances can be evaluated based on the slope of the transmission coefficient curves with the element change. While for the non-identical triple-layer unit-cell, there are no specific curves that describe the variations in the transmission coefficient vs. the change in the dimensions of two independent conductor layers. But, we can estimate the transmission coefficient error of each conductor layer separately through the slope of the transmission coefficient curves of the single-layer vs. the element change. In this section, we aim to demonstrate the change in transmission coefficient of the non-identical triple-layer unit-cell due to a small change in the transmission phase, and hence change in the corresponding transmission magnitude, of each single-layer that may arise due to manufacturing tolerances.

Figure 5.17 demonstrates an example of a non-identical triple-layer unit-cell with transmission phase of the first and third layers equal to $\left(\angle S_{21}|_{layer1} = \angle S_{21}|_{layer3} = 71°\right)$ and a transmission phase of the second layer equals to $\left(\angle S_{21}|_{layer2} = 10°\right)$ The transmission coefficient of this unit-cell is presented by a blue dot in the polar diagram of Fig. 5.17. This unit-cell has high transmission magnitude value with transmission phase around $0°$. However, with a small change of

$\pm 1°$ in the transmission phase of any layer, the transmission coefficient of the unit-cell changes drastically, as shown in Fig. 5.17. All possible changes in the transmission coefficient of this unit-cell due to phase tolerances of $\pm 1°$ and $\pm 2°$ in each layer are presented by the red arcs in Figs. 5.18a,b, respectively. These results clarify that this unit-cell is very sensitive to manufacturing errors.

Figure 5.19 presents two other cases of non-identical unit-cells but with smaller transmission magnitude values. We can notice that, when the transmission magnitude of the unit-cell with non-identical layers decreases, the sensitivity to manufacturing tolerances decreases and the phase differences between the non-identical layers decreases too.

According to the study of more non-identical unit-cells with different transmission coefficient values, we can clarify the causes that increase the unit-cell sensitivity to manufacturing tolerances as follows.

(a) The sensitivity to manufacturing tolerances increases when the phase difference between the first/third layers $(\angle S_{21}|_{layer1} = \angle S_{21}|_{layer3})$ and the second layer $(\angle S_{21}|_{layer2})$ increases.

(b) The required phase range around $0°$ of the non-identical unit-cell, as shown in Fig. 5.14, is very close to the resonance of the element, which have a very small transmission coefficient when all layers are identical. This region is very sensitive to manufacturing tolerances.

Improvements in Sensitivity to Manufacturing Precision
Based on these observations, the ideal target of Fig. 5.14 is very difficult to be implemented using the unit-cell configuration of Fig. 5.12 with $\varepsilon_r = 1$ and, $\beta L_d = 90°$, due to the high sensitivity to manufacturing precision around transmission phase of $0°$. However, a practical target that is shown in Fig. 5.20 can be considered as follows.

(a) A transmission phase range from $30°$ to $330°$ can be obtained using identical conductor layers $(\angle S_{21}|_{layer1} = \angle S_{21}|_{layer3} = \angle S_{21}|_{layer2})$, with transmission magnitude better than or equal to -2 dB.

- A transmission phase range from $-30°$ to $30°$ can be obtained using non-identical conductor layers $(\angle S_{21}|_{layer1} = \angle S_{21}|_{layer3} \neq \angle S_{21}|_{layer2})$, such that each connductor layer is individually designed, but keeping the first and third layers identical, and maintaining the transmission magnitude equal to -2 dB, as shown in Fig. 5.20.

To further improve the sensitivity of the non-identical triple-layer unit-cell to manufacturing precision, we should avoid the transmission phase range around $0°$, which is close to the resonance region. Figure 5.21 presents the case when the electrical separation between layers decreases to $\beta L_d = 30°$. It shows that the region around $0°$ can be covered by an identical triple-layer unit-cell with transmission magnitude close to 1 (0 dB) except in a small region the transmission magnitude decreases to -2 dB. This unit-cell of identical layers has reduction in the transmission magnitude for transmission phase values around $110°$, as shown in Fig. 5.21. By studying

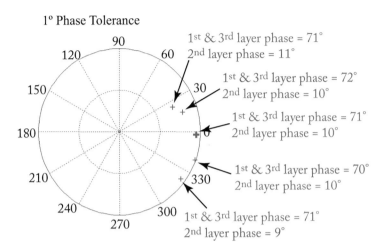

Figure 5.17: Sensitivity of a non-identical, triple-layer unit-cell with $\varepsilon_r = 1$ and $\beta L_d = 90°$ to phase variations of $\pm 1°$ in the conductor layers.

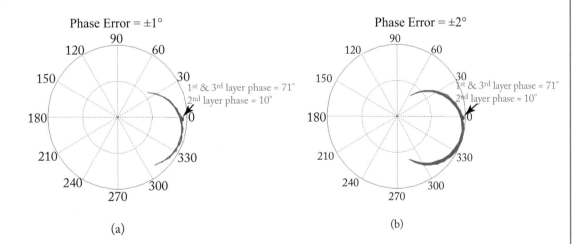

Figure 5.18: Sensitivity of a high transmission coefficient, non-identical triple-layer unit-cell with $\varepsilon_r = 1$ and $\beta L_d = 90°$ to phase tolerance in each layer equal to: (a) $\pm 1°$ and (b) $\pm 2°$.

Transmission Magnitude of The Unit-Cell = -1dB

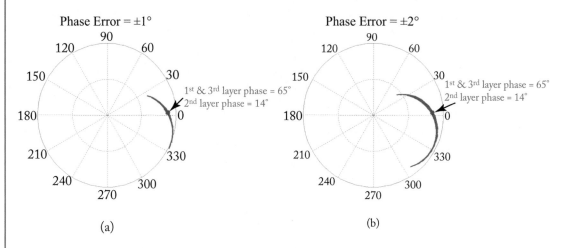

(a)   (b)

Transmission Magnitude of The Unit-Cell = -2dB

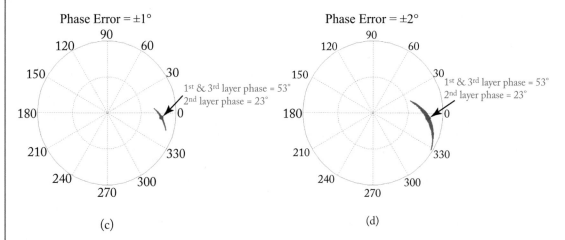

(c)   (d)

Figure 5.19: Sensitivity of a non-identical, triple-layer unit-cell with $\varepsilon_r = 1$ and $\beta L_d = 90°$ and for different transmission magnitude values to phase tolerance in each layer.

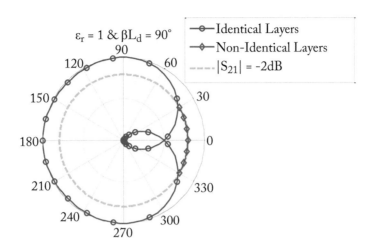

Figure 5.20: Practical full transmission phase range of 360° for a triple-layer FSS configuration using a combination of both identical and non-identical layers with air-gap separation between layers ($\varepsilon_r = 1$) and electrical thickness of $\beta L_d = 90°$.

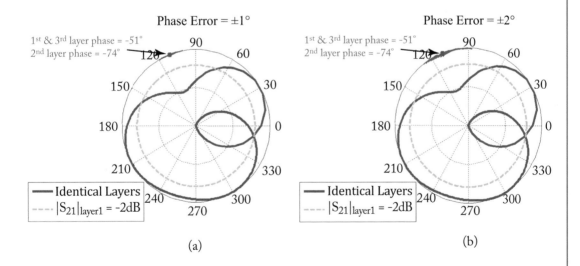

Figure 5.21: Sensitivity of a high transmission coefficient, non-identical, triple-layer unit-cell with $\varepsilon_r = 1$ and $\beta L_d = 30°$ to phase tolerance in each layer equal to: (a) $\pm 1°$ and (b) $\pm 2°$.

the non-identical unit-cell configuration of Fig. 5.12 with $\varepsilon_r = 1$ and $\beta L_d = 30°$ in this phase range, we noticed that the unit-cell is less sensitive to the phase error of the individual layers, as shown in Fig. 5.21, compared to the case when of $\beta L_d = 90°$ of Fig. 5.18. Accordingly, by using a triple-layer configuration with $\varepsilon_r = 1$ and $\beta L_d = 30°$, a full transmission phase of $360°$ can be obtained as follows.

(a) Elements with identical layers $(\angle S_{21}|_{layer1} = \angle S_{21}|_{layer3} = \angle S_{21}|_{layer2})$ is used for transmission phase that ranges from $-165°$ to $35°$.

(b) Elements with non-identical layers $(\angle S_{21}|_{layer1} = \angle S_{21}|_{layer3} \neq \angle S_{21}|_{layer2})$ is used for transmission phase that ranges from $35°$ to $195°$, as shown in Fig. 5.22.

5.3 DOUBLE-LAYER UNIT-CELLS

A double-layer unit-cell separated by either an air-gap or a dielectric substrate was discussed in Section 3.2. Based on the results of Fig. 3.16 and Table 3.2, a maximum transmission phase range of $228.5°$ with transmission coefficient reduction of -3 dB can be obtained using an air gap $(\varepsilon_r = 1)$ between the two conductor layers with electrical separation equal to $\beta L_d = 155°$. Actually, this phase range can also be obtained with an electrical separation between layers equal to $\beta L_d = 205°$, as shown in Fig. 5.23. We cannotice that the transmission coefficient curves in polar diagrams for these two electrical separations are in opposite directions along the horizontal axis (compare Fig. 5.23b to Fig. 3.16).

Accordingly, a full transmission phase range of $360°$ can be obtained when combining the phase ranges of these two unit-cells, which have layer separations of $155°$ and $205°$, respectively. But in this case, we have to take into account the difference in electrical thickness between the two unit-cells $(\Delta = 205° - 155° = 50°)$ when calculating the transmission coefficients of the unit-cell that have smaller thickness, as shown in Fig. 5.24. The electrical difference Δ is actually a phase shift that is added to the transmission phase of the unit-cell, which leads to the rotation of the transmission coefficient curve by $50°$ in the clockwise direction, as shown in Fig. 5.24b.

Figure 5.25 demonstrates the required transmission phase ranges that could be obtained from each unit-cell, such that a full transmission phase range of $360°$ is achieved with transmission magnitude better than or equal to -3 dB by combining the two phase ranges. A double-layer unit-cell with electrical thickness of $\beta L_d = 205°$ could be used for the transmission phase that ranges from $-25°$ to $190°$, while the other unit-cell with electrical thickness of $\beta L_d = 155°$ could be used for the remaining transmission phase range from $190°$ to $335°$.

In order to validate these analytical results, two double-layer unit-cells for the configurations of Figs. 5.23 and 5.24 are simulated using CST Microwave Studio software [27]. The element geometry is the double square loop element of Fig. 3.6 at 11.3 GHz. Figure 5.26a,b are the simulated transmission magnitude and phase, respectively, of the two unit-cells vs. the outer square loop side length. Figure 5.26c depicts the transmission magnitude and phase of the two

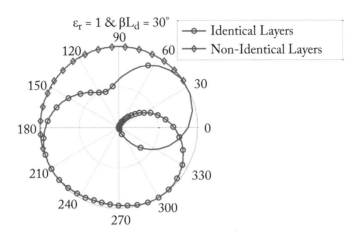

Figure 5.22: Full transmission phase range of 360° for a triple-layer FSS configuration using a combination of both identical and non-identical layers with air-gap separation between layers ($\varepsilon_r = 1$) and electrical thickness of $\beta L_d = 30°$.

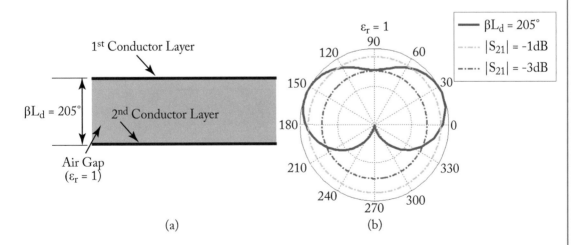

Figure 5.23: Transmission coefficients of a double-layer unit-cell using air gap ($\varepsilon_r = 1$) with electrical separation equal to $\beta L_d = 205°$: (a) the double-layer configuration and (b) transmission coefficient in a polar diagram.

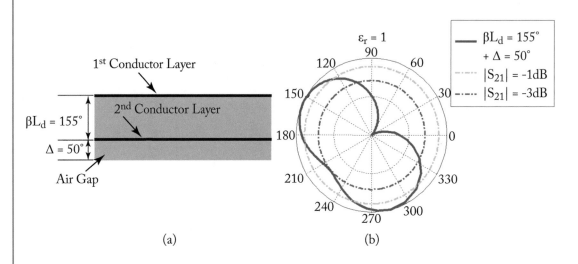

Figure 5.24: Transmission coefficients of a double-layer unit-cell using air gap ($\varepsilon_r = 1$) with electrical separation between layers equal to $\beta L_d = 155°$ and taking into account the difference in electrical thickness of ($\Delta = 50°$) with the unit-cell of Fig. 5.23: (a) the double-layer configuration and (b) transmission coefficient in a polar diagram.

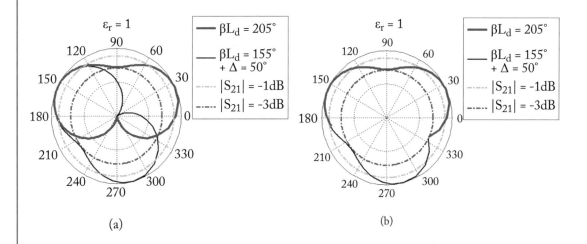

Figure 5.25: Transmission coefficients of the two double-layer unit-cells of Figs. 5.23 and 5.24 in a single polar plot: (a) complete curves and (b) required transmission phase range from each unit-cell.

unit-cells in a single polar diagram. A good agreement between full wave simulation and analytical results can be observed. The results validate the idea of obtaining a full transmission phase range of 360° when combining the transmission phase ranges of two double-layer unit-cells, which have different thicknesses.

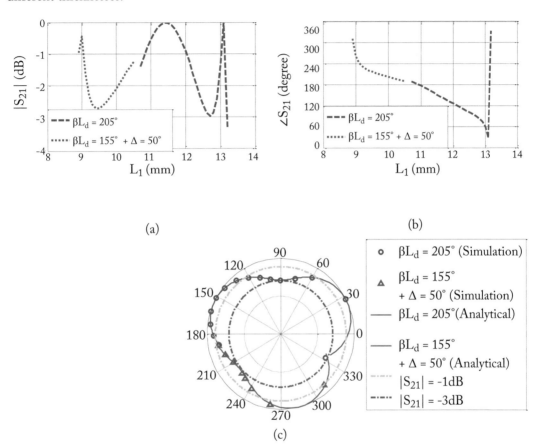

Figure 5.26: Transmission coefficients of the two double-layer unit-cells of Figs. 5.23 and 5.24 using double square loop element: (a) transmission magnitude, (b) transmission phase, and (c) polar plot.

For further clarification of the design concept, Fig. 5.27, demonstrate a circular aperture transmitarray antenna using the double square loop elements in two groups of double-layer configurations. The first layer is a common layer for the two unit-cell groups. The second layer includes only the elements of the unit-cell group that has smaller thickness ($\beta L_d = 155°$), while the third layer includes only the elements of the unit-cell group that has larger thickness ($\beta L_d = 205°$). In practice, each layer of conductor elements could be mounted on a thin substrate and the separation between layers could be a foam layer with $\varepsilon_r \approx 1$. However, this design concept has a challenge for the accurate use of periodic boundary conditions in the numerical simulations of two combined unit-cells with different thicknesses.

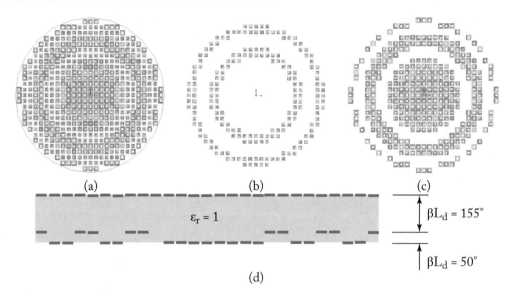

(a)   (b)   (c)

$\varepsilon_r = 1$

$\beta L_d = 155°$

$\beta L_d = 50°$

(d)

Figure 5.27: Transmitarray antenna design using two groups of double-layer unit-cells, which have different thicknesses: (a) mask of the first layer, (b) mask of the second layer, (c) mask of the third layer, and (d) side-view cut.

CHAPTER 6

Wideband Transmitarray Antennas

Transmitarray antennas have several advantages compared to lens antennas such as low profile, lightweight, low fabrication cost, and versatile functionalities through the individual control of transmitarray elements. However, transmitarray antennas have narrow bandwidth due to the narrow band limitation of the transmitarray elements and the differential spatial phase delay resulting from the different lengths from the feed to each element on the transmitarray aperture.

Various efforts have been made to increase the bandwidth of transmitarray antennas. One approach involves using multiple identical layers of relatively wideband elements [4, 39, 40]. A proposed wideband transmitarray antenna using 6 layers of Jerusalem cross elements at 30 GHz has been presented in [39]. A quad-layer transmitarray antenna using dual-resonant double square loops achieves 7.5% 1 dB gain-bandwidth at 30.25 GHz, with aperture efficiency of 35.6% [6]. In [40], a triple-layer transmitarray antenna achieves a 1 dB gain bandwidth of 9% with 30% aperture efficiency at 11.3 GHz using spiral dipole elements, which has been presented in Section 5.1.

Another approach involves using receiver-transmitter design [16–19]. A reconfigurable 1-bit transmitarray antenna achieves 15.8% 3-dB gain bandwidth at 9.8 GHz using PIN diodes, with aperture efficiency of 15.4% that is calculated based on maximum gain and aperture dimensions available in [16, 19]. In [17] 7.1% and 7.6% 1 dB gain bandwidths with aperture efficiencies of 17% and 12.9%, respectively, have been achieved using 1-bit transmitarrays at 60 GHz. In [18] a stacked patch reconfigurable transmitarray element using varactor diodes had been studied, which achieves 10% 3 dB fractional bandwidth with 400° phase range and an insertion loss varying between 2 dB and 5 dB. But this high insertion loss values will lead to low aperture efficiency.

There are other types of wideband planar lenses used for focusing the electromagnetic waves. Periodic sub-wavelength metamatrials [21–23] and band-pass frequency selective surfaces [7, 43] are the most common methods used to design this type of planar lenses. It is noted that most of the ideas being made to increase the bandwidth of transmitarray antennas are at the expenses of the aperture efficiency and design complexity.

This chapter presents a detailed study on the transmission magnitude and phase of transmitarray elements as a function of frequency, aiming to improve the transmitarray antenna bandwidth. We demonstrate a new design methodology for improving the bandwidth of transmitarray antennas through the control of the transmission phase range and the optimization of the phase distribution on the transmitarray aperture. The novelty of this work focuses on aperture distribution synthesis to enhance the bandwidth, which is general for any element shape, while most of the other designs focus on using wideband elements. It is important to note that the proposed techniques do not preclude implementation of wideband elements in transmitarray designs, and a combination of multiple broadband techniques would implicitly yield a better bandwidth performance.

In order to validate this technique, two quad-layer transmitarray prototypes using double square loop elements have been designed, fabricated, and tested at Ku-band. The transmission phase distribution is optimized for both antennas, while they have different transmission phase ranges. The results show wideband performances of 9.8% and 11.7% for 1 dB gain, with aperture efficiencies of 50% and 47%, respectively, at 13.5 GHz.

6.1   BANDWIDTH ANALYSIS OF A TRANSMITARRAY USING QUAD-LAYER DOUBLE SQUARE LOOP ELEMENTS

6.1.1   UNIT-CELL PROPERTY

In this study, we select a double square loop element with four identical layers as a reference to analyze the bandwidth characteristics. The unit-cell periodicity of $P \approx \lambda_0/2 = 11.1$ mm, where λ_0 is the free space wavelength at 13.5 GHz. The geometrical model of the element along with the design parameters is shown in Fig. 6.1. The elements are printed on a dielectric substrate with thickness $T = 0.5$ mm and permittivity $\varepsilon_r = 2.574$. The separation between layers is equal to $H = 5$ mm, such that the total separation between two layers is close to a quarter wavelength ($H + T \approx \lambda_0/4 = 5.56$ mm) [2].

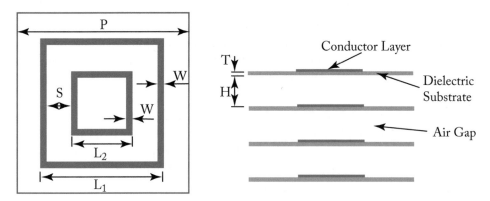

Figure 6.1: The quad-layer unit-cell configuration of a double square loop element: (a) top view and (b) side view.

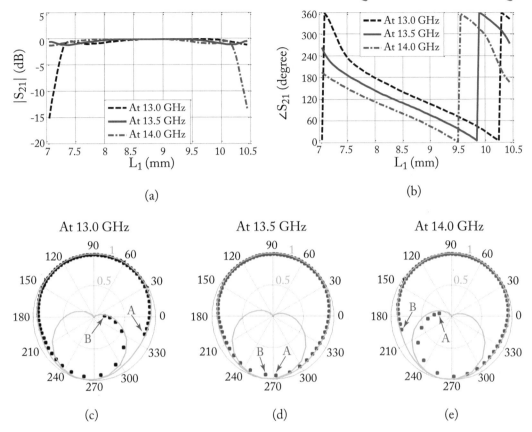

Figure 6.2: Transmission coefficients at different frequencies: (a) magnitudes, (b) phases, (c) polar plot at 13.0 GHz, (d) polar plot at 13.5 GHz, and (e) polar plot at 14 GHz.

The unit-cell simulations were carried out using the commercial software CST Microwave Studio [27]. Periodic boundaries were imposed on the four sides of the unit-cell to simulate an infinite array of elements. Absorbing boundaries are considered on the top and bottom surfaces of the unit-cell, and a normal incidence plane wave is used to illuminate the unit-cell element. Parametric studies were performed to determine the separation between the two loops (S) and loop width (W) with varying the outer loop length L_1. The optimum dimensions were determined to be $S = 0.22L_1$ and $W = 0.4$ mm.

Figure 6.2a,b, depict the transmission magnitudes and phases of the unit-cell element as a function of the outer loop length L_1 and at different frequencies. It is worthwhile to present these results in polar diagrams as a function of L_1, as shown in Fig. 6.2c to e. The polar plot magnitude represents the transmission magnitude, i.e., $|S_{21}|$ and the angle represents the transmission phase, i.e., $\angle S_{21}$. By varying the outer loop length, L_1 from 7.05–14.45 mm (corresponding to points B and A), a full phase range of 360° with a transmission magnitude equal to or better than -1.2 dB at the center frequency of 13.5 GHz can be achieved with this element.

However, at lower frequencies (such as 13 GHz), the transmission coefficient curve on the polar diagram rotates counterclockwise and follows the theoretical curve (green curve), as shown in Fig. 6.2c. For example, point A rotates from the 270° location in Fig. 6.2d to the 340° location in Fig. 6.2c. This leads to a decrease in the transmission magnitude (see Fig. 6.2a) and an increase in the slope of the transmission phase (see Fig. 6.2b) around point B with a small value of L_1.

Similarly, at higher frequencies (such as 14 GHz), the transmission coefficient curve rotates clockwise on the polar diagram, as shown in Fig. 6.2e, which consequently leads to a decrease in the transmission magnitude (see Fig. 6.2a) and an increase in the slope of the transmission phase (see Fig. 6.2b) around point A with a large value of L_1.

In summary, as the frequency changes, the transmission coefficients of the elements change, as shown in Fig. 6.2c,e, leading to both phase error and magnitude loss, which ultimately results in a reduction of antenna gain at off-center frequencies.

For more clarification, the transmission magnitudes and phases versus frequency for differ-ent values of L_1 are presented in Fig. 6.3. We cannotice the magnitude reduction and the change in slope of the phase, which occur simultaneously at low frequencies for small values of L_1 (e.g., $L_1 = 7.5$ mm), and occur at high frequencies for large values of L_1 (e.g., $L_1 = 10$ mm).

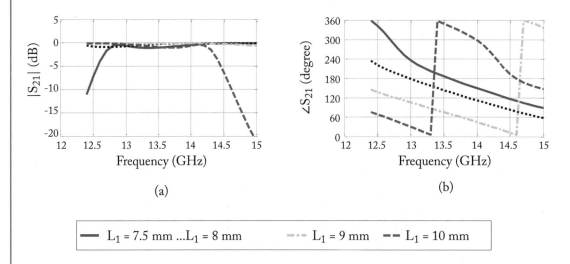

(a)

(b)

——— $L_1 = 7.5$ mm ... $L_1 = 8$ mm   - - - $L_1 = 9$ mm   – – $L_1 = 10$ mm

Figure 6.3: Transmission coefficients vs. frequency for different values of L_1: (a) magnitudes and (b) phases.

6.1.2 BANDWIDTH PERFORMANCE OF TRANSMITARRAY

The phase distribution of the transmitarray aperture was discussed in Section 2.1, and the transmission phase ψ_i for the ith element was given in Equation (2.1) as:

$$\psi_i = k\left(R_i - \vec{r}_i \cdot \hat{r}_o\right) + \psi_0, \tag{6.1}$$

where k is the propagation constant in free space R_i, is the distance from the feed source to the ith element, $\vec{r}_i$ is the ith element position vector, and $\hat{r}_o$ is the main beam unit vector, as shown in Fig. 6.4. The phase constant ψ_0 can be added to all elements of the array. Once the phase of the ith element is determined, the corresponding outer loop length, L_1, can be obtained from Fig. 6.2b.

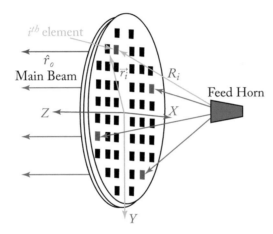

Figure 6.4: Geometry of a printed transmitarray antenna.

To demonstrate the effects of both phase error and magnitude loss on the transmitarray bandwidth, a quad-layer transmitarray antenna using the double square loop elements in Fig. 6.1 is designed. The transmitarray has a circular aperture with a diameter of $14.5\lambda_0 = 32.19$ cm, and an F/D ratio of 0.95, where λ_0 is the free space wavelength at 13.5 GHz. The transmitarray aperture has 621 elements. The feed horn is vertically polarized (along the y-direction) with a gain equal to 16.3 dB and half-power beamwidths (HPBW) of 30° at 13.5 GHz. The feed horn pattern is approximately modeled as $\cos^q(\theta)$ with $q = 9.25$, which corresponds to an edge taper of -10.2 dB on the transmitarray aperture. Moreover, the phase constant ψ_0 is selected deliberately for optimum performances (It will be discussed in details in the next section).

Using the transmission magnitude and phase properties shown in Fig. 6.2, the array theory [28] is used to calculate the antenna gain as a function of frequency at five different cases, as follows.

- In the ideal case, the transmission magnitude is equal 1 (0 dB) and the transmission phase changes with frequency according to Equation (6.1). Thus, the element phase change cancels the spatial phase effect at all frequencies.

- In the case of differential spatial phase effect, we assume constant transmission phase along the frequency band under consideration. Thus, phase error occurs only due to the change of path lengths from the feed to each element on the transmitarray aperture with the change of frequency.

- To consider the case of only phase error effect, the element magnitude is selected to be equal to 1 (0 dB), while the phase properties of Fig. 6.2 are considered.

- Similarly, the case of only magnitude loss effect is demonstrated by considering only the magnitude properties of Fig. 6.2, while the element phase changes with frequency according to Equation (6.1).

- The practical case is to consider both magnitude and phase properties of the element.

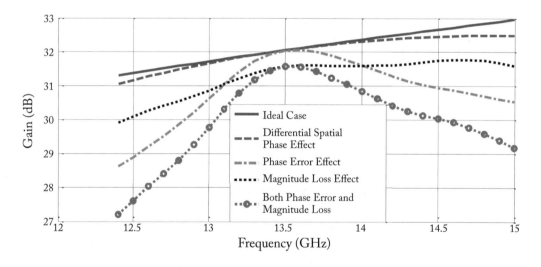

Figure 6.5: Effects of element phase error and element loss on the transmitarray antenna gain.

At the center frequency (13.5 GHz), the phase error is almost zero because the element achieves the full phase range of 360° when varying its dimensions as shown in Fig. 6.2d. However, the phase error limits the antenna bandwidth due to the differential spatial phase delay and the change in the slope of the element phase vs. element dimensions that occurs at off-center frequencies, as mentioned in Section 6.1.1. The element magnitude loss shows less impact on bandwidth limitation compared to the phase error effect. However, it reduces the antenna gain. This gain reduction increases at off-center frequencies, as discussed in Section 6.1.1. For example, at the center frequency 13.5 GHz, the gain reduction is 0.47 dB, while at the off-center frequencies, the gain reductions are 0.85 dB at 13 GHz and 0.77 dB at 14 GHz.

6.2 BANDWIDTH PERFORMANCE WITH DIFFERENT REFERENCE PHASES AT THE APERTURE CENTER

In this section, we study the effect of the phase constant, ψ_0, on the bandwidth of the transmitarray antenna. For this phase analysis, we consider the reference point to be the center of the aperture, which has a transmission phase value of ψ_c. The optimum phase constant is then determined by studying all possible values of phase in one full cycle (360°).

Several quad-layer transmitarray antennas using the same double square loop elements of Fig. 6.1 are studied here. The transmitarrays have the same configuration as that presented in the previous section (such as aperture shape and diameter, number of elements, and feed characteristics). They differ only in the aperture phase constant, ψ_0. To illustrate the phase constant effect, Fig. 6.6 demonstrates antenna gain for two of these transmitarrays as a function of frequency and also compared with the ideal case. The corresponding phase at the aperture center, ψ_c, for these two arrays presented here are 10° and 270°, respectively. It is observed that different phase constants will lead to different bandwidth results.

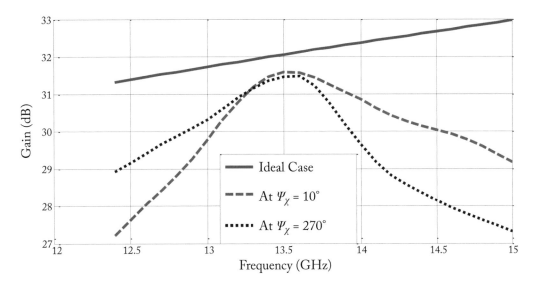

Figure 6.6: Calculated gain for different phase values at the aperture center.

For better interpretation of these results, it is advantageous to observe the transmission magnitude on the aperture, since the impact of each element on the overall performance of the array also depends on the illumination of that particular element. In most cases, such as in the study here, the feed antenna is pointing to the geometrical center of the array, thus the center elements have a stronger illumination and contribute more to the overall performance of the array.

It can be seen from Fig. 6.7b,e, that at the center frequency of 13.5 GHz, the transmission magnitudes of all elements are better than -1.2 dB, corresponding to Fig. 6.2d. Thus, the change in the phase constant ψ_0 (accordingly the center phase ψ_c) does not have much effect on the antenna gain at 13.5 GHz, as shown in Fig. 6.6.

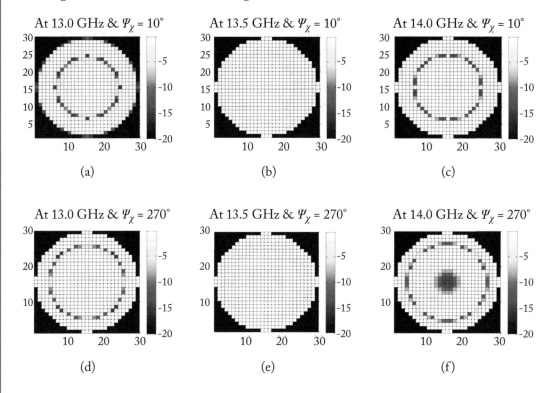

Figure 6.7: Transmission magnitudes on the transmitarray aperture in dB with two different phase values at the aperture center ψ_c and at three different frequencies.

At the lower frequency of 13.0 GHz and referring to Fig. 6.2c,d, we can expect the best selection of the aperture center phase is $\psi_c = 270°$ at the center frequency (equivalent to 340° at 13.0 GHz), which is represented by point A in Fig. 6.2c,d. The selection of this aperture center phase places the elements with smaller transmission magnitudes (which implicitly include phase errors) at the farthest positions away from the geometrical center of the aperture, as shown in Fig. 6.7d. Thus, the aperture phase at the center $\psi_c = 270°$ is considered the best selection along a range of lower frequencies. This provides a solid explanation on why the antenna achieves a higher gain at the lower frequencies when ψ_c is set to 270°, as shown in Fig. 6.6.

On the other hand, at the higher frequency of 14.0 GHz and referring to Fig. 6.2d,e, we can expect the worst selection of the aperture center phase is $\psi_c = 270°$ (element resonates at 14.0 GHz), which is represented by point A in Fig. 6.2d,e. It leads the elements that have smaller magnitude (which implicitly include phase errors) to start at the aperture center, as shown in Fig. 6.7f. This clarifies the reason of having low antenna gain values at the higher frequencies when $\psi_c = 270°$, as shown in Fig. 6.6. It is important to clarify that using the element's transmission magnitude response to optimize the aperture phase distribution is implicitly led to the decrease of the effect of the element's phase error, which is associated with the transmission magnitude reduction at off-center frequencies.

Through a parametric study, the case of $\psi_c = 10°$ shows the best element distributions and widest bandwidth compared to the other cases. Table 6.1 summarizes the performances of five transmitarrays, which have different phase values at the center of the aperture.

Table 6.1: Comparison of transmitarray antennas differ in the phase values at the center of the aperture

Aperture Center Phase Ψ_c	Antenna Gain at 13.5 GHz	Aperture Efficiency at 13.5 GHz	1 dB Gain Bandwidth
0°	31.57 dB	68.37%	6.9%
10°	31.58 dB	68.49%	7.0%
90°	31.46 dB	66.67%	5.5%
180°	31.34 dB	64.85%	5.1%
270°	31.46 dB	66.66%	5.9%

6.3   PROPER SELECTION OF ELEMENT PHASE RANGE FOR IMPROVEMENT OF TRANSMITARRAY BANDWIDTH

Based on the results of Fig. 6.2, it is clear that selecting a range of outer loop dimensions, L_1, which can achieve the full phase range of 360° at a certain frequency, will result in transmission coefficient variation at other frequencies. In particular, large variation (both magnitude reduction and phase slope change) occurs for elements with dimensions that correspond to a transmission phase around 270° at the center frequency, which can best be observed in the polar diagrams of Fig. 6.2. Accordingly, in order to minimize the effect of these elements across the frequency band of interest, one could avoid using elements that have transmission phases around 270°.

To study the feasibility of this technique, four new quad-layer transmitarray antennas are designed using the double square loop elements in Fig. 6.1. The transmitarrays have the same configuration as those studied in the previous sections. Also, the aperture phase at the center for all four antennas is set to $\psi_c = 10°$. The four transmitarrays differ only in the transmission phase ranges of their elements at the center frequency, which are 360°, 300°, 240°, and 180°, respectively. We note that these phase ranges have been carefully selected to avoid using specific elements that have transmission phases around 270°. For better demonstration of the phase range selection, Fig. 6.8a,b, present in polar diagram the phase ranges for two cases of these four transmitarrays. The polar diagrams of Fig. 6.8c,d, present the transmission coefficients for the case of limited phase range of 240° at lower and higher frequencies, respectively. The figures demonstrate avoiding of those elements with low transmission coefficients at off-center frequencies in comparison with the case of full phase range of Fig. 6.2c,e, respectively. Figure 6.9 depicts the calculated gain vs. frequency of these two transmitarrays presented here. A summary of the performances of the arrays is also given in Table 6.2.

Comparison of the gain bandwidths in Fig. 6.9 shows that as expected, limiting the transmission phase range by avoiding elements with transmission phases around 270°, increases the antenna gain bandwidth. In particular, since elements with poor transmission magnitude at extreme frequencies are not used in these arrays, the antenna gain at these extreme frequencies increase, which ultimately results in an overall increase of antenna gain bandwidth.

Table 6.2: Comparison of four transmitarray antennas differ in the element transmission phase ranges

Transmission Phase Range	Antenna Gain at 13.5 GHz	Aperture Efficiency at 13.5 GHz	1 dB Gain Bandwidth
360°	31.58 dB	68.49%	7.0%
300°	31.51 dB	67.39%	7.8%
240°	31.12 dB	61.65%	10.3%
180°	29.48 dB	42.21%	15.5%

It is also important to note that exclusion of these elements and consequently using a phase range less than a full cycle result in some reduction of antenna gain at the center frequency. The transmitarray bandwidth can be increased at the expense of some compromise in gain and aperture efficiency at the center frequency. The influence of the element phase range on gain bandwidth and aperture efficiency of the transmitarrays is depicted in Fig. 6.10.

Improvement of the transmitarray bandwidth through the control of the transmission phase range does not dispense with the use of the optimization process that was discussed in Section 6.2. Because although the limitation of the transmission phase range around 270° avoids the reduction in transmission magnitude of the transmitarray elements along a band of frequencies, it increases the transmission phase error due to phase truncation. This in turn leads to some reduction in the antenna gain especially at the center frequency. Therefore, optimizing the phase distribution in this case aims to keep the region of truncated phase (around 270°) away as much as possible from

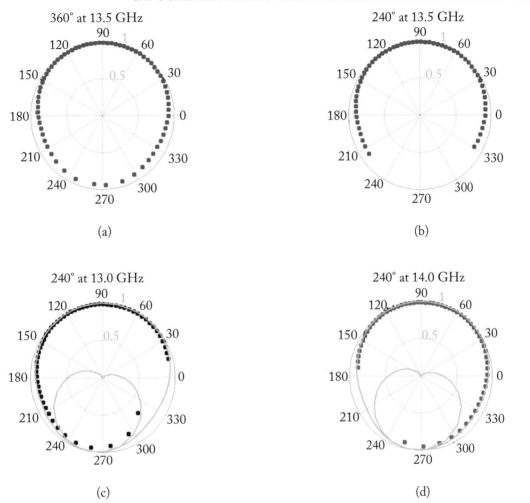

Figure 6.8: Two different transmission phase ranges: (a) 360° at 13.5 GHz, (b) 240° at 13.5 GHz, (c) 240° at 13.0 GHz, and (d) 240° at 14.0 GHz.

the aperture center. This in turn leads to minimize the impact of the truncated phase to reduce the antenna gain.

6.4   COMPARISON BETWEEN DIFFERENT ELEMENT SHAPES

The relation between magnitude and phase of an element in a multilayer frequency selective surface (M-FSS) is determined by the number of layers, the substrate material, and the separation

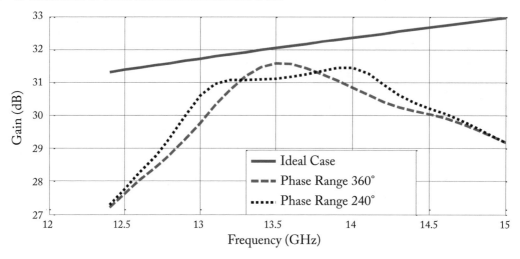

Figure 6.9: Transmitarray calculated gain for different transmission phase ranges.

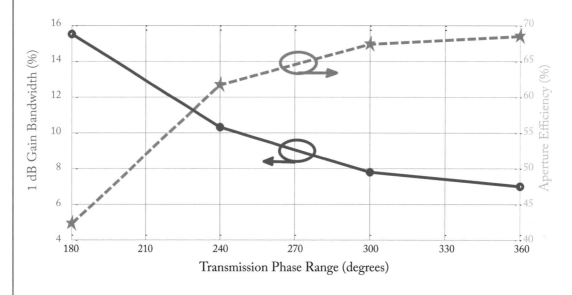

Figure 6.10: Aperture efficiency and 1 dB gain bandwitdth vs. transmission phase range.

between layers, regardless of the element shape. The change in the element transmission magnitude is generally a function of the transmission phase values. Accordingly, the proposed design technique to improve the bandwidth of transmitarray antennas is feasible for general element shapes. However, for an element with a specific shape, its transmission coefficient changes distinctly with respect to its dimensions. Thus, the bandwidth values that can be obtained differ from one element shape to another.

For further clarification, the bandwidth characteristics of another two different elements are studied. The elements, as shown in Fig. 6.11, are the double four-legged loaded element (DFLL) [37, 43–45] and the Jerusalem crosselement. The unit-cell configurations of these elements, such as number of layers, unit-cell periodicity, substrate material, and layer separation, are the same as that presented in Section 6.1 The two unit-cells were simulated using the CST Microwave Studio [27]. Parametric studies were performed to determine the optimum element dimensions. For the DFLL element, the separation between loops $S = 0.2L$, $d = 0.2L$, and the width $W = 0.3$ mm. While for the Jerusalem cross element, the side length $L_s = 0.7L$, and the width $W = 0.5$ mm. The full phase range of $360°$ is achieved by varying the element dimension L, where L varies from 7.25 mm to 10.45 mm for the DFLL element, and from 5.35 mm to 10.25 mm for the Jerusalem cross element.

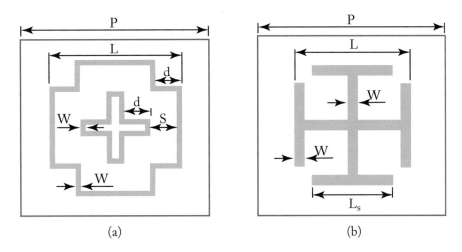

(a)   (b)

Figure 6.11: (a) DFLL element and (b) Jerusalem cross element.

For a comprehensive comparison between the three element shapes, eight new quad-layer transmitarray antennas are designed with the same configuration as those presented in the previous sections. The phase at the aperture center for all eight antennas are selected equal to $\psi_c = 10°$. Four of these transmitarrays used the DFLL elements in Fig. 6.11a with different transmission phase ranges. The other four transmitarrays used the Jerusalem cross elements in Fig. 6.11b, which also differ in the transmission phase range. These phase ranges have been carefully selected to avoid using specific elements that have transmission phases around $270°$.

Figure 6.12 demonstrates the bandwidth performance of the three element shapes with the influence of the element phase range. We can notice that the three curves are almost parallel, which indicates that the bandwidth improvement using the proposed technique is feasible for general element shapes. However, the bandwidth values that can be obtained differ from one element shape to another. Regarding the three element shapes under consideration, the double square loop element has the widest bandwidth performance. Meanwhile, the gain and the corresponding aperture efficiency values are almost same for the three elements at the center frequency, which is not shown here for brevity.

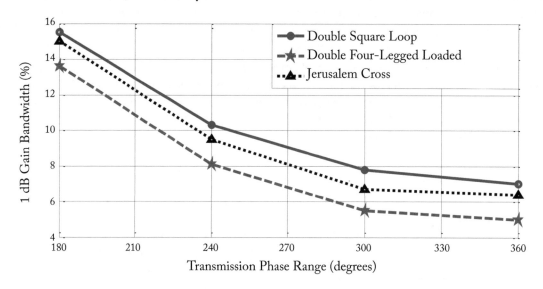

Figure 6.12: Bandwidth of 1 dB gain vs. transmission phase range of three different element shapes.

6.5 PROTOTYPE FABRICATION AND MEASUREMENTS

To validate the proposed bandwidth improvement method, two quad-layer transmitarray antennas using double square loop elements have been fabricated and tested. The two prototypes are the two cases with phase ranges of 360° and 240° that were presented in Table 6.2. For both designs, an optimum transmission phase distribution is selected for the array, which corresponds to a reference phase at the aperture center of $\psi_c = 10°$. The design parameters of Antenna 1 that has full phase range and Antenna 2 that has limited phase range are summarized in Table 6.3.

The two antennas are identical in every parameter except the range of outer loop length, L_1, which are selected based on the designated element phase range. Figure 6.13a illustrates two transmitarray masks with the difference in dimensions for some elements of the two antennas. Figure 6.13b shows the elements that are different in the two antennas due to the difference in the range of dimension L_1. The dots represent the elements that are same in the two antennas, and the "x" symbols represent the elements that are different. The elements that are different are 140 elements out of 621 total elements. The elements of each layer are printed on a dielectric substrate. Plastic screws and spacers are used to maintain an equal separation between layers.

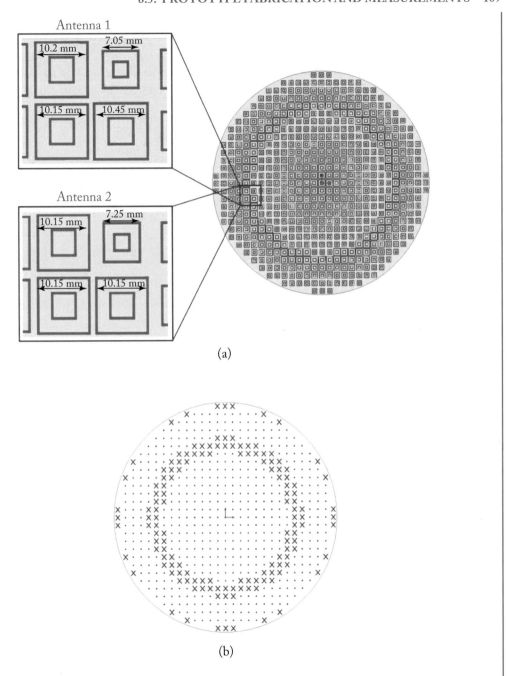

Figure 6.13: (a) Transmitarray mask with the difference in dimensions for some elements of the two antennas and (b) elements that are different in the two antennas, as represented by the "*x*" symbol.

Table 6.3: Design configurations of the two transmitarray prototypes

	Antenna 1 360° Phase Range	Antenna 2 240° Phase Range
Number of conductor layers	4	
Substrate thickness (T)	0.55 mm	
Substrate permittivity (ε_r)	2.574	
Layer separation (H)	5 mm	
Unit-cell periodicity (P)	11.1 mm	
Number of elements	621	
Aperture diameter	32.19	
Feed to diameter ratio (F/D)	0.95	
Feed q factor at 13.5 GHz	9.25	
Range of outer loop length (L_1)	(7.05–10.45) mm	(7.25–10.15) mm
Loop widths (W)	0.4 mm	
Loop separation (S)	0.22 L_1	

The fabricated prototypes are tested using the NSI planar near-field measurement system. A photo of the test setup is shown in Fig. 6.14. Figure 6.15a,b, show the measured gain patterns of Antenna 1 and Antenna 2, respectively, at the center frequency of 13.5 GHz. The simulated co-polarized gain patterns, calculated using array theory [28], are also shown for comparison. At 13.5 GHz, the measured gain of Antenna 1 and Antenna 2 is 30.22 dB and 29.95 dB, respectively. This corresponds to aperture efficiencies of 50% and 47%, respectively. The half-power beamwidths (HPBWs) for both antennas are same and equal to 4.9° and 5° in the E- and H-plane, respectively. The measured side-lobe levels (SLL) of Antenna 1 and Antenna 2 are −22 dB and −20 dB, respectively. The cross-polarized levels of both antennas are equal to −30 dB.

Figure 6.16 demonstrates the calculated and measured gain versus frequency of the two antennas, which confirms the proposed methodology to increase the bandwidth of the transmitarray antennas. Optimization of the reference phase at the aperture center ψ_c for both antennas improves the transmitarray bandwidth. Moreover, it can be noticed that reducing the transmission phase range by avoiding elements with phases around 270° in Antenna 2, leads to the increase of antenna gain at higher and lower frequencies compared to the case of having full phase range in Antenna 1, but with a slight decrease in antenna gain at the center frequency. This in turn increases the transmitarray bandwidth. We also noticed that the measurements show slow decline in the gain at low frequencies compared to the theoretical results, leading to wider bands than expectations. In summary, the measurements show wideband performances of 9.8% and 11.7%

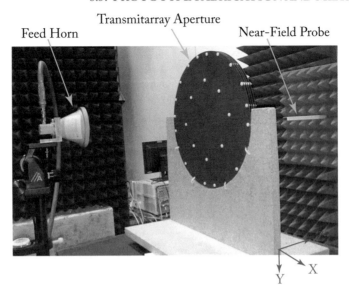

Figure 6.14: Measurement setup of a transmitarray antenna using the NSI planar new-field system.

for 1 dB gain for Antenna 1 and Antenna 2, respectively. Table 6.4 summarizes the measurement results.

Table 6.4: Measurement results of the two transmitarray prototypes

	Antenna 1	Antenna 2
Gain at 13.5 GHz	30.22 dB	29.95 dB
Aperture efficiency at 13.5 GHz	50%	47%
1 dB gain bandwidth	9.8%	11.7%
HPBW at 13.5 GHz (E-plane, H-plane)	4.9°, 5°	4.9°, 5°
Sidelobe level at 13.5 GHz	-22 dB	-20 dB
Cross-polarized level at 13.5 GHz	-30 dB	-30 dB

Meanwhile, it is noticed that the measured gains are about 1.2 dB lower than simulation results at the center frequency. We consider these discrepancies are due to the fabrication errors, feed alignments, and approximations of the simulation model.

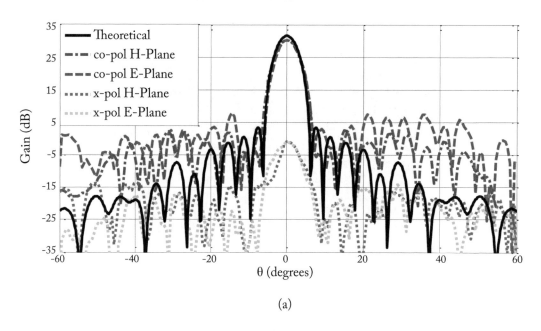

(a)

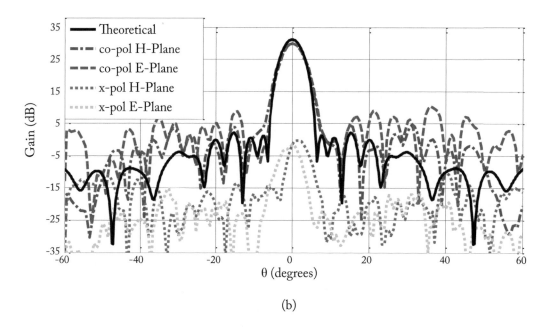

(b)

Figure 6.15: Measured and simulated radiation patterns at 13.5 GHz: (a) Antenna 1 with full phase range and (b) Antenna 2 with limited phase range.

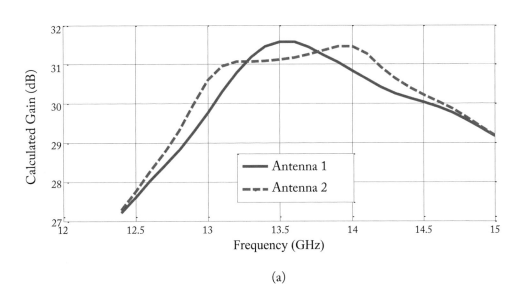

(a)

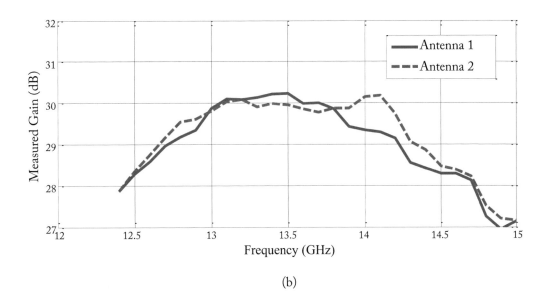

(b)

Figure 6.16: Gains vs. frequency of the two antennas: (a) theoretical and (b) measurement.

CHAPTER 7

Single-feed Multi-beam Transmitarrays

In comparison with a dielectric lens antenna, a prominent advantage of a transmitarray antenna is the capability to individually control each element phase on the array aperture, which provides a simple mechanism to synthesize the array and achieve desired radiation performance. Owing to this feature, diversified radiation patterns can be achieved with transmitarray antennas in a simple fashion [46].

High-gain multi-beam antennas typically rely on reflectors or lenses with feed-horn clusters or large phased arrays. The primary drawbacks of these systems are typically cost, weight, and volume, particularly for space applications. Similar to reflectarray antennas [47, 48], multiple simultaneous beams can be achieved with a transmitarray antenna by using a single feed at no additional cost, with the added advantages of low-mass and low-profile features. In comparison with multi-beam reflectarray antennas, the main advantage of multi-beam transmitarray systems is that they don't exhibit feed blockage, which removes any constraint on the beam directions.

In this chapter we study the radiation characteristics of single-feed transmitarray antennas with simultaneous multiple beams, through case studies of quad-beam designs. A powerful global search algorithm, particle swarm optimization, is implemented to synthesize the phase of the transmitarray antenna elements. Various pattern masks and fitness functions are studied for multi-beam designs, and a Ku-band quad-beam transmitarray antenna using quad-layer double square loop elements is demonstrated with 24.77 dB gain for each beam. Due to the nature of the design, the synthesized arrays do not exhibit a progressive phase on the aperture as traditional single-beam designs, and as a consequence, significant differences in dimensions occur between each element with its surrounding neighbor elements. Accordingly, the periodic approximation in the unit-cell simulation, which consider all elements are identical, are carefully studied to investigate its impact on the transmission coefficients of the unit-cell element. Moreover, the effects of the element phase error and magnitude loss on the radiation patterns are also studied.

7.1 DESIGN METHODOLOGIES FOR SINGLE-FEED MULTI-BEAM TRANSMITARRAY ANTENNAS

In transmitarray antennas, the array element taper is fixed by the feed properties and the element locations; however, the elements of a transmitarray antenna have the flexibility to achieve any value of phase shift. Utilizing this direct control of phase shift for every element, the phase distribution on the array aperture can be synthesized to achieve simultaneous multiple beams using a single feed. In other words, designing a multi-beam transmitarray is basically a phase-only array synthesis problem.

In general, two different synthesis approaches are available for single-feed multi-beam array designs: direct analytical solutions or optimization methods. While analytical solutions are typically simple to implement, recent studies [47] have shown that the achievable performance of these methods is not satisfactory in many cases. Hence, it is necessary to implement some form of optimization routine to achieve good radiation characteristics. These optimization approaches however require a robust search algorithm and an efficient pattern computation engine in order to synthesize the phase of a transmitarray antenna with several hundreds of elements. In terms of the search algorithms, while local search methods, such as alternating projection method can achieve a good performance in some cases [46], a global search method is preferable since it can avoid local minima traps in non-convex optimization problems such as asymmetric multi-beam designs [48].

In this study we use the powerful particle swarm optimization (PSO) method [49] for synthesizing the aperture phase distribution of transmitarray antennas for multi-beam operation. In the first step, far-field pattern masks are defined based on the design requirements. These masks are circular contours in the angular space that are defined in the direction of each beam [47, 48]. Fitness functions are then defined for the optimizations, which control the main beam performance and side-lobe level of the array. The pattern computation in this synthesis process is conducted efficiently using an in-house code that is based on the array theory formulation with spectral transformations for computational speedup [28].

7.2 DESIGN OF KU-BAND SINGLE-FEED QUAD-BEAM TRANSMITARRAY ANTENNAS

To demonstrate the feasibility of the design approach proposed here for single-feed multi-beam transmitarray designs, we studied a symmetric quad-beam system with 50° elevation separation between the beams, i.e., the beams point at ($\theta_{1,2,3,4} = 25°, \varphi_1 = 0°, \varphi_2 = 90°, \varphi_3 = 180°, \varphi_4 = 270°$). The antenna has a circular aperture with 648 elements and is designed for the operating frequency of 13.5 GHz. The unit-cell size is 11.1 mm, and the phasing elements are the quad-layer double square loops (QLDSL) in Fig. 7.1 with a layer separation of 5 mm, which can achieve a 360° phase range with a transmission magnitude better than −1.2 dB at 13.5 GHz, as shown in Fig. 7.2. The feed antenna is a linearly polarized corrugated conical horn with a q value of 9.25 at 13.5 GHz, which is placed at a distance of 275 mm from the antenna aperture.

Three different designs of quad-beam transmitarrays are studied here. As discussed earlier, the radiation pattern is controlled by defining a far-field pattern mask. Two different masks are considered: a constant side-lobe of −30 dB (Design 1 of Fig. 7.3), and a tapered side-lobe of −25 dB to −40 dB (Design 2 of Fig. 7.4). A two-term fitness function is defined which evaluates

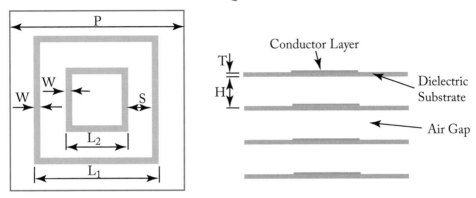

Figure 7.1: The quad-layer unit-cell configuration of a double square loop element: (a) top view and (b) side view.

the radiation performance of the array in terms of the peak gain for each beam and the side lobe levels in the entire angular space based on the mask requirements [48]. The fitness function to be minimized is:

$$Cost = W_1 \sum_{\substack{(u,v) \notin \text{mainbeam} \\ \text{and} \, |F(u,v)| > M_U(u,v)}} \sum (|F(u,v)| - M_U(u,v))^2$$

$$+ W_2 \sum_{\substack{(u,v) \in \text{mainbeam} \\ \text{and} \, |F(u,v)| < M_L(u,v)}} \sum (|F(u,v)| - M_L(u,v))^2. \tag{7.1}$$

For the optimization, a swarm population of 150 particles is selected for the PSO and two symmetry planes are defined to reduce the size of the solution hyperspace. Numerical studies showed that for the tapered mask, the penalty for the main-beam fitness term had to be increased to achieve a better performance, thus a third design with double penalty for main-beam fitness was also studied (Design 3 of Fig. 7.5). The double penalty relates to the weights associated with the terms in the fitness function in Equation (7.1), where in the main beam area, $W_2 = 2W_1$.

A summary of the gain performances of these three designs is given in Table 7.1. In summary, all three quad-beam transmitarrays studied here achieved a good multi-beam performance, which demonstrates the effectiveness of the proposed approach for multi-beam designs. Among the three designs, Design 3 achieves the highest gain, with the best overall radiation performance in the entire angular space.

Ideal elements, by definition, do not exhibit any losses or phase errors. In other words, the ideal element has transmission phase matching the exact phase requirement on the aperture, and a transmission magnitude equal to 1 (0 dB). However, in the case of QLDSL element, the transmission magnitude and phase properties of Fig. 7.2 are considered in the gain calculation. Using the ideal element case clarifies the gain performance of the three synthesis design approaches.

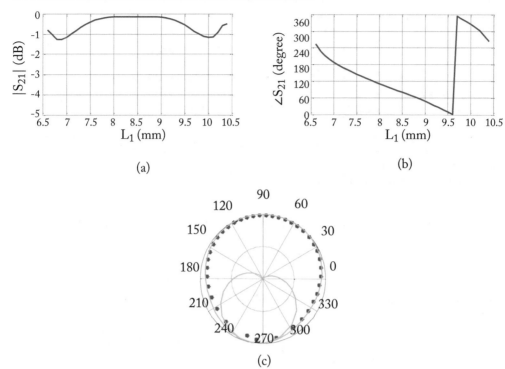

Figure 7.2: Transmission coefficient of the double square loop element with normal incidence at 13.5 GHz: (a) transmission magnitude, (b) transmission phase, and (c) polar plot.

However, in order to calculate the expected gain when using the QLDSL element, the transmission magnitude and phase of the QLDSL element should be considered.

Table 7.1: Comparison of different designs of single-feed quad-beam transmitarray antennas

Transmitarray	Gain (ideal elements)	Gain (QLDSL elements)
Design 1	25.19 dB	24.56 dB
Design 2	24.95 dB	24.15 dB
Design 3	25.35 dB	24.77 dB

7.3   PROTOTYPE FABRICATION AND MEASUREMENTS

Design 3 of the optimized quad-beam transmitarray is then fabricated and measured at the design frequency of 13.5 GHz. The mask and the photograph of one layer of the fabricated array with

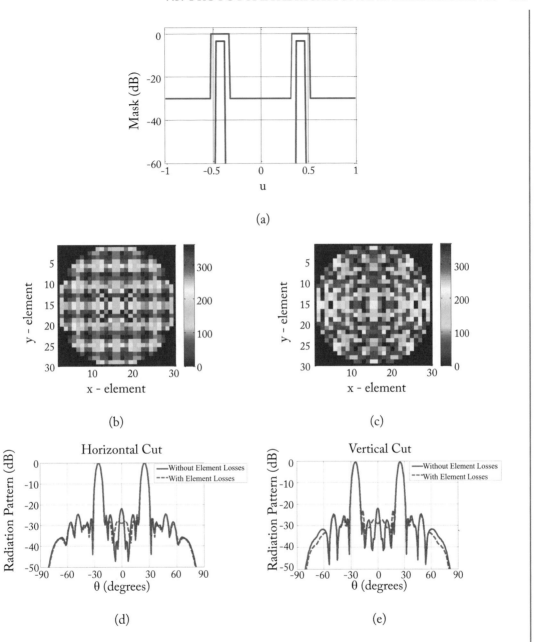

Figure 7.3: Design 1: A constant side-lobe mask: (a) ideal pattern, (b) optimized phase distribution without space delay term, (c) actual phase distribution including space delay term, (d) radiation patterns along horizontal cut, and (e) radiation patterns along vertical cut.

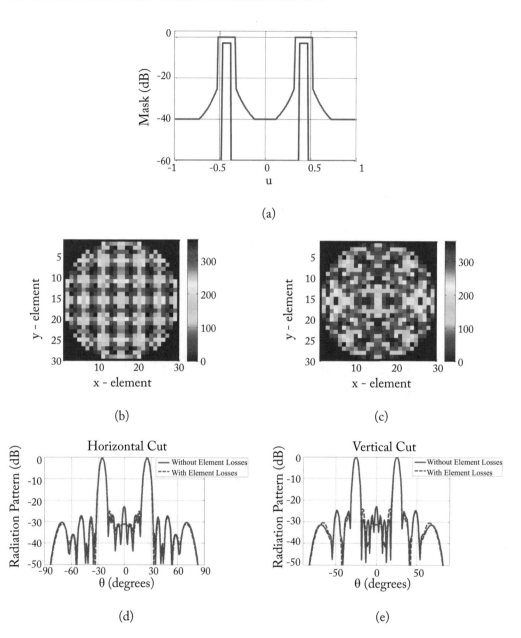

Figure 7.4: Design 2: A tapered side-lobe mask: (a) ideal pattern, (b) optimized phase distribution without space delay term, (c) actual phase distribution including space delay term, (d) radiation patterns along horizontal cut, and (e) radiation patterns along vertical cut.

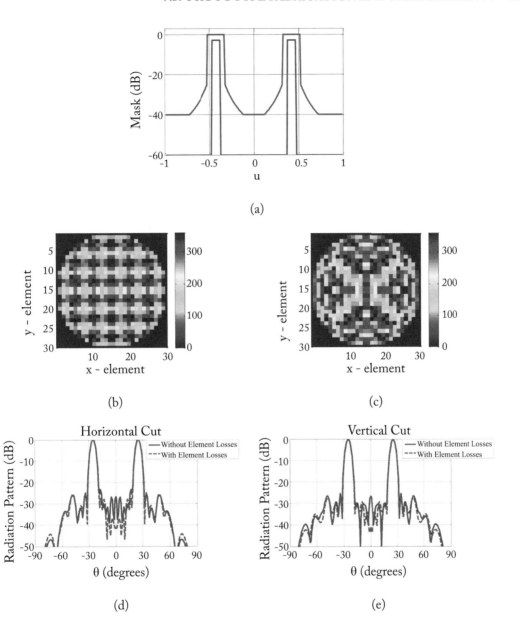

Figure 7.5: Design 3: A tapered side-lobe mask and double penalty in main beam regions: (a) ideal pattern, (b) optimized phase distribution without space delay term, (c) actual phase distribution including space delay term, (d) radiation patterns along horizontal cut, and (e) radiation patterns along vertical cut.

648 quad-layer double square loop (QLDSL) elements are shown in Fig. 7.6. The fabricated prototype is tested using the NSI planar near-field measurement system with 52×52 inch sampling plane (166×166 samples), at a distance of 11 inch from the near-field probe, which leads to a truncation level of -19.41 dB. A photo of the test setup is shown in Fig. 7.7.

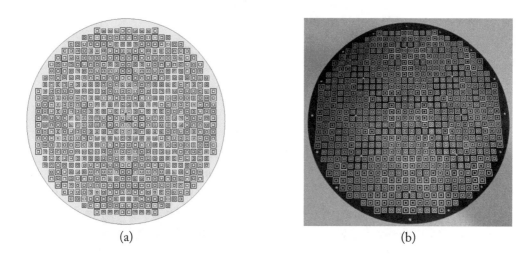

(a)   (b)

Figure 7.6: One layer of the fabricated quad-beam transmitarray prototype: (a) mask and (b) photograph.

Figure 7.7: Near-field measurement setup of the single-feed quad-beam transmitarray antenna.

The near-field and far-field radiation patterns are demonstrated in Figs. 7.8 and 7.9, respectively, which illustrate quad-beam with high gain performances. The four beams are located at elevation angle $\theta_{1,2,3,4} = 25°$, except a 1° shift in one beam, and azimuth angles $\varphi_1 = 0°$, $\varphi_2 = 90°$, $\varphi_3 = 180°$, and $\varphi_4 = 270°$. The measured gain of the two beams along the yz-plane are the same and are equal 23.81 dB, and those along the xz-plane are equal 22.33 dB and 22.66 dB. The side lobe and cross polarization levels are less than -14 dB and -30 dB, respectively.

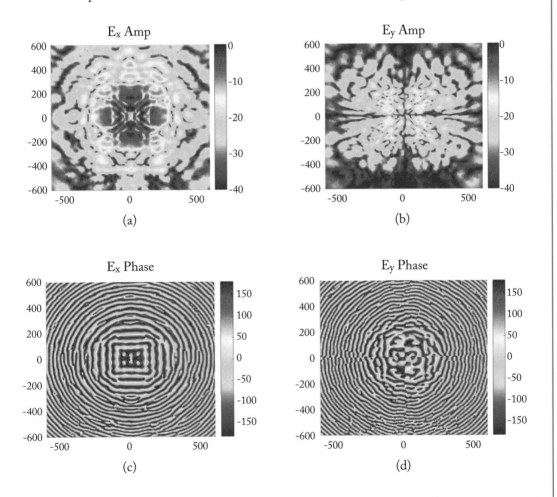

Figure 7.8: Near-field patterns: (a) co-pol amplitude, (b) x-pol magnitude, (c) co-pol phase, and (d) x-pol phase.

A gain reduction of 1.15 dB and 1.48 dB was observed for the two beams along the xz-plane compared with the other two beams along the yz-plane. This reduction is mainly due to polarization effect. Moreover, the side lobe levels between the beams are considered high com-

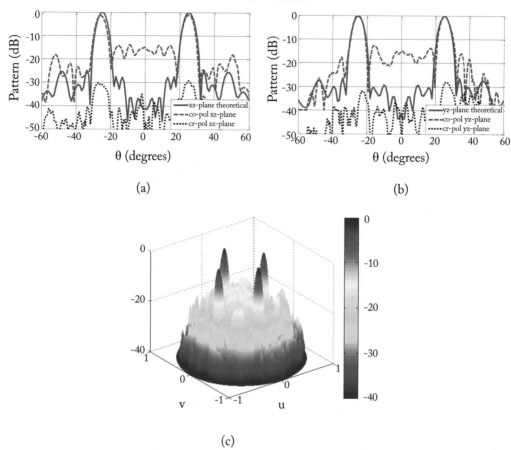

Figure 7.9: Far-field patterns at 13.5 GHz: (a) xz-plane, (b) yz-plane, and (c) 3-D pattern.

pared with the design requirements. We think the main causes of the increase of the side lobe levels are as follows (see also Section 2.6).

1. Approximations in the unit-cell analysis.

   a. Normal incidence approximation, which neglects the oblique incidence effect.

   b. Infinite array approximation, which assumes that each element is surrounded by an infinite number of similar elements with the neglect of the effects of the variations in dimensions of neighbor elements.

2. Fabrication tolerances.

Accordingly, we have carefully studied the oblique incidence effect and the impact of the variations in dimensions of the neighbor elements on the transmission coefficients of the unit-cell element.

Moreover, the impacts of the element phase error and magnitude loss on the radiation patterns are also studied.

7.4   TRANSMITARRAY APPROXIMATION AND PERFORMANCE DISCUSSIONS

7.4.1   OBLIQUE INCIDENCE EFFECT OF THE UNIT-CELL ELEMENT

It is worthy to study the behavior of the unit-cell element under oblique incidence and clarify whether the normal incidence approximation of the unit-cell analysis could have significant effect on the measured radiation pattern. Figure 7.10 depicts the variations in the transmission magnitude and phase of the quad-layer double square loop element at different oblique incidence angles and for y-polarized incidence signal. The parameters θ and ϕ are the elevation and azimuth angles, respectively, of the incidence wave.

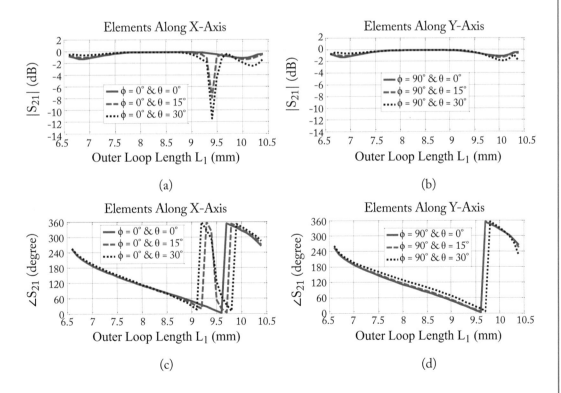

Figure 7.10: Transmission coefficients of the double square loop element vs. element dimension L_1 under different incident angles: (a) magnitude of elements along x-axis, (b) magnitude of elements along y-axis, (c) phase of elements along x-axis, and (d) phase of elements along y-axis.

The results show that there is almost no significant change in transmission coefficients with oblique incidence. Only for elements that have dimensions of $L_1 = 9.4$ mm and are along the x-axis ($\phi = 0°$), these elements could have significant magnitude reduction with oblique incidence angles. However, the fabricated prototype under consideration has only four elements with dimensions of $L_1 = 9.4$ mm. Moreover, these four elements are located at $\phi = 23°$ (not along the x-axis) and at the aperture edge, thus low contribution on the antenna patterns. Accordingly, we do not consider the normal incidence approximation of the unit-cell analysis is the main causes of the increase in the measured side lobe levels.

7.4.2   VARIATIONS IN DIMENSIONS OF NEIGHBORING ELEMENTS

The optimization process for the multi-beam transmitarray design led to a non-uniform phase distribution of the transmitarray aperture. This in turn led to significant difference in dimensions between each element with its surrounding neighbor elements (see Fig. 7.6). Accordingly, the periodic approximations in the unit-cell simulation, which consider all elements are identical, could lead to notable error in the transmission coefficient values. In order to investigate the accuracy of the unit-cell element approximations, a large unit-cell consists of nine neighbor elements is studied. The dimensions of the nine elements of this large unit-cell are different, which are selected from real samples of the designed prototype antenna. The results of the large unit-cell are then compared with those of the conventional unit-cell.

Three different cases, as shown in Fig. 7.11, are simulated using CST Microwave Studio software [27]. Due to the symmetry of the transmitarray mask, these three large unit-cells are also located in the other three quadrants of the mask. The dimensions L_1 of the center element for the three cases are 7.20 mm, 7.75 mm, and 8.85 mm, respectively. The dimensions of the other neighbor elements are selected according to their actual dimensions in the designed quad-beam transmitarray prototype. The dimensions of the neighboring elements are summarized in Table 7.2.

Table 7.2: Dimensions L_1 of the neighboring elements for the three cases of the large unit-cell

Large Unit-Cell	Dimensions L_1 of the Eight Neighbor Elements (mm)	Mean Error	Standard Deviation
Case 1, center cell = 7.20 mm	10.05, 7.25, 10.35, 8.55, 9.45, 10.3, 7.8, 7.05	1.65 mm	1.37
Case 2, center cell = 7.75 mm	8.2, 7.15, 7.1, 7.15, 9.45, 8.2, 7.2, 10.35	0.35 mm	1.23
Case 3, center cell = 8.85 mm	7.9, 9.05, 9.65, 9.3, 8.6, 7.1, 7.9, 8.25	0.18 mm	0.85

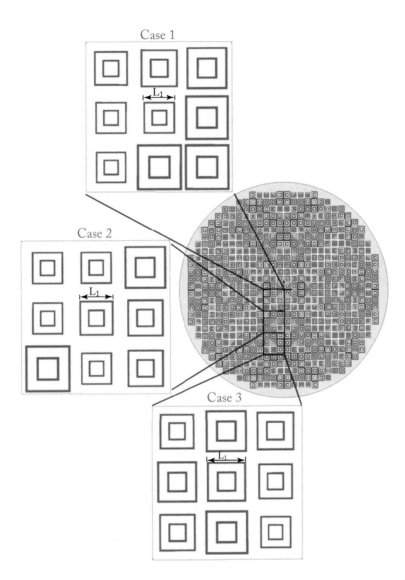

Figure 7.11: Large unti-cell analysis.

The transmission coefficients of the three cases are compared with those of the conventional unit-cell element in Fig. 7.12. Due to the asymmetry of the large unit-cell, the transmission coefficients of both perpendicular (TE) and parallel (TM) modes are considered. It can be noticed that Case 1 and Case 2 have both large phase error and magnitude loss when compared with the conventional unit-cell element. However, Case 3 has almost no significant change in the transmission coefficient values, because the range of variations in dimensions of the nine elements in Case 3 is smaller than those in Case 1 and Case 2. As given in Table 7.2, Case 3 has the smallest mean error and standard deviation. These results show that the variations in dimensions of neighboring elements lead to both phase error and magnitude loss, which is the main reason for the discrepancy between the measured and simulated patterns.

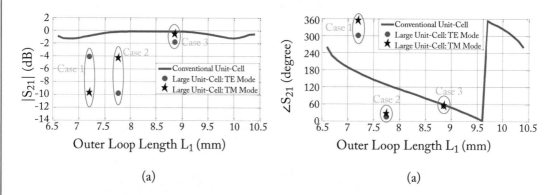

Figure 7.12: Transmission coefficients of the large unit-cell compared with the conventional unit-cell: (a) transmission magnitude and (b) transmission phase.

7.4.3   PHASE ERROR AND MAGNITUDE LOSS EFFECT ON THE RADIATION PATTERNS

This section aims to study the effects of both the phase error and the magnitude loss of the unit-cell element on the radiation patterns of the quad-beam transmitarray prototype. For phase error analysis, a random phase is added to the actual phase of each element using normal distribution with mean value of 0°. The standard deviation of this normal distribution ranges from 10° to 70°. For each standard deviation value, 20 trials of radiation patterns are demonstrated, as shown in Fig. 7.13. Besides, the average of these 20 trials is also presented in. The new transmission phase ψ_{i_new} of the ith element is calculated as:

$$\psi_{i\_new} = \psi_i + rand\,(\mu = 0, \sigma)\,, \tag{7.2}$$

where ψ_i is the actual phase of the ith element, and $rand(\mu = 0, \sigma)$ is a random phase error using normal distribution with mean of $\mu = 0$ and standard deviation equal to σ (degree).

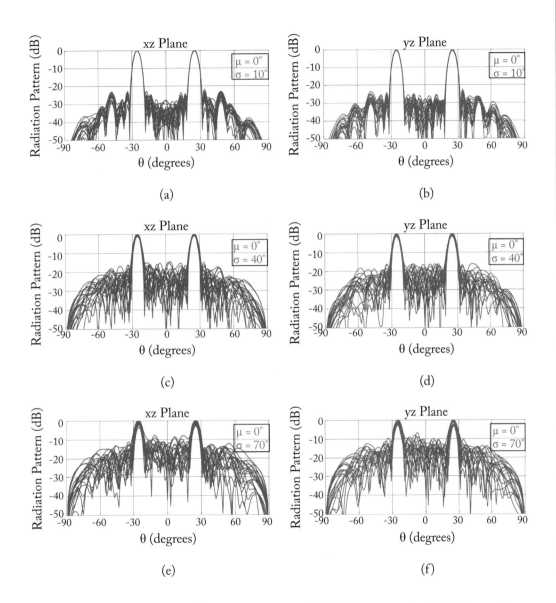

Figure 7.13: Radiation patterns of 20 transmitarray trials for different standard deviations of the random phase error distribution.

In the same way for the magnitude loss analysis, a random magnitude loss is added to the actual magnitude of each element using normal distribution with mean value of 0 dB and with different standard deviation values that ranges from -5 dB to -15 dB. Because the magnitude loss led to a reduction in the transmission magnitude, the random magnitude losses must be negative values in dB. The new transmission magnitude $|T_{i\_new}|$ of the ith element is calculated as:

$$|T_{i\_new}|\ (\text{dB}) = |T_i|\ (\text{dB}) - abs\,[rand\,(\mu = 0, \sigma)]\,, \qquad (7.3)$$

where $|T_i|$ is the actual transmission magnitude of the ith element in dB, and $rand(\mu = 0, \sigma)$ is a random magnitude error using normal distribution with mean of $\mu = 0$ and standard deviation equal to σ (dB). Similar to the phase error analysis, for each standard deviation value, 20 trials of radiation patterns are demonstrated in Fig. 7.15, and the average of these 20 trials is presented in Fig. 7.16.

These results reveal that while both phase error and magnitude loss of the transmitarray elements have little effect on the direction of the main beams, they significantly increase the side-lobe levels. In particular, the side-lobes in the area between the four beams increases by 20 dB with a random phase error with a $40°$ standard deviation.

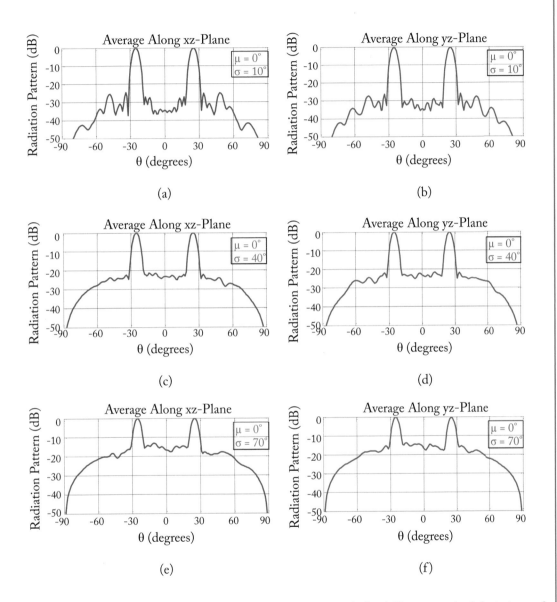

Figure 7.14: Average radiation patterns of 20 transmitarray trials for different standard deviations of the random phase error distribution.

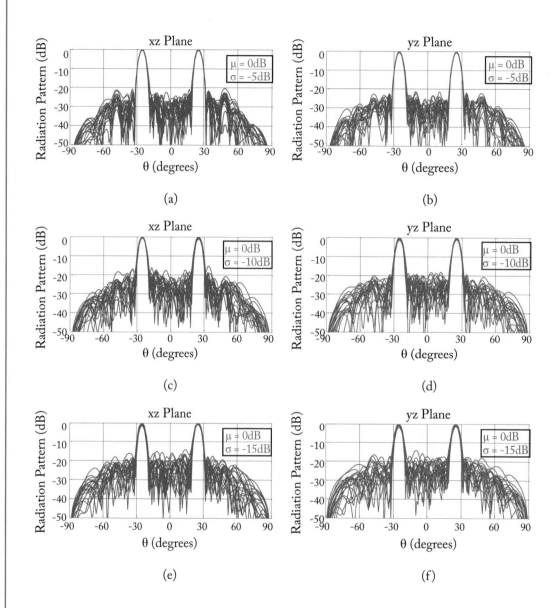

Figure 7.15: Radiation patterns of 20 transmitarray trials for different standard deviations of the random magnitude loss distribution.

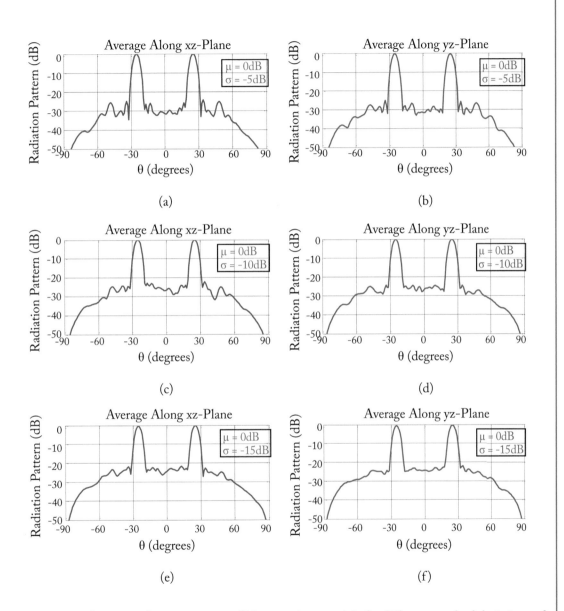

Figure 7.16: Average radiation patterns of 20 transmitarray trials for different standard deviations of the random magnitude loss distribution.

CHAPTER 8

Conclusions

Transmitarray antennas combine many favorable features of optical lens and microstrip array antennas, leading to a low profile and low mass design with high radiation efficiency and versatile radiation performance. Compared to reflectarray antennas, the transmitarrays encounter great challenges for both magnitude and phase control of the elements as well as the bandwidth limitation.

8.1 CONTRIBUTIONS OF THIS BOOK

New methodologies and novel designs have been presented, and some exciting progress was obtained accordingly. The primary contributions of this book are as follows.

- The transmission performance of multilayer frequency selective surfaces (M-FSS) has been carefully studied for transmitarray designs. The transmission phase limits of the M-FSS structure have been revealed, which is general for arbitrary FSS geometries. The maximum transmission phase range has been determined according to the number of layers, substrate permittivity, and separation between conductor layers. It is revealed that the -1 dB transmission phase limits are 54°, 170°, 308°, and full 360° for single-, double-, triple-, and quad-layer FSS consisting of identical layers, respectively. These analytical limits are generally applicable, independent from the selection of a specific element shape.

- A quad-layer transmitarray antenna using cross-slot elements has been designed, fabricated, and tested at X-band. This design has a novelty in using slot-type elements with no dielectric substrate, which has two main advantages. The first advantage is its suitability for space applications, because compared with the dielectric substrates, the conductor layers can bear extreme temperature changes in outer space. The second advantage is the cost reduction, because there is no need to use high-performance microwave substrate. Moreover, a detailed analysis of the transmitarray antenna considering the oblique incidence angles and the feed polarization conditions is performed, where their impacts on the antenna gain and the radiation patterns are clearly demonstrated.

- In order to reduce the transmitarray design complexity and cost, three different methods are investigated to design triple-layer transmitarray antennas. The overall performance is maintained with full transmission coefficient phase range of 360° while avoiding the reduction of the element transmission coefficient magnitude. Based on this analysis, a novel high gain broadband triple-layer transmitarray antenna using spiral-dipole elements has been designed, fabricated, and tested at X-band. The spiral-dipole element has a large range of variation in the element dimensions, creating a more linear slope for the phase, which makes the design less sensitive to manufacturing error. The element phase distribution on the transmitarray aperture is optimized to decrease the loss due to the elements having small transmission magnitudes, leading to an average element loss as low as 0.49 dB. This design achieves a measured gain of 28.9 dB at 11.3 GHz, aperture efficiency of 30%, and a broadband of 9% at 1 dB-gain and 19.4% at 3 dB-gain.

• New design methodologies are proposed to improve the bandwidth of transmitarray antennas. Variation in the transmission coefficient of transmitarray elements as a function of frequency is first studied. This study clarifies that for quad-layer transmitarrays, elements with transmission phases around 270° suffer from the deterioration in both transmission magnitude and phase with the change of frequency. This in turn limits the transmitarray bandwidth. Accordingly, to increase the bandwidth of a transmitarray antenna, a phase truncation is performed in the element selection routine, which avoids certain elements around this frequency-sensitive phase region. Moreover, an optimization of the phase distribution of transmitarray elements is carried out. It aims to keep the elements, which have either low transmission magnitude at off-center frequencies or have phase truncation at the center frequency, away as much as possible from the aperture center. This in turn minimizes the impact of these elements in reducing the gain along a band of frequencies, and hence increases the antenna gain bandwidth. The proposed design methodology has been validated through the fabrication and testing of two quad-layer transmitarray antennas at Ku band. The measurements show high gains of 30.22 dB and 29.95 dB at 13.5 GHz, leading to aperture efficiencies of 50% and 47%, respectively, and wideband performances of 9.8% and 11.7%, respectively, for 1 dB gain bandwidth.

• The feasibility of designing single-feed transmitarray antennas with simultaneous multiple beams is investigated through case studies of quad-beam designs. The powerful global search algorithm, particle swarm optimization, is implemented to synthesize the phases of the transmitarray antenna elements. Various pattern masks and fitness functions are studied for multi-beam designs. A novel Ku-band single-feed quad-beam transmitarray antenna with 50° elevation separation between the beams are designed, fabricated and tested. Moreover, the periodic unit cell approximation has been discussed, and the impact of the element phase error on the radiation patterns of the quad-beam transmitarray antenna has been demonstrated.

8.2   FUTURE WORK

The main focus of this book is on demonstrating critical analysis and new methodologies for the design of transmitarray antennas using multilayer frequency selective surfaces approach (M-FSS).

Meanwhile, a variety of possible topics are still open. Although our analysis confirms multilayer configuration is required to achieve full phase range of 360°, one may proceed in the analysis of multilayer transmitarrays but with very small layer separation, where strong coupling of higher order modes occurs. This in turn will lead to a very low profile antenna design with less sensitive to oblique incidence angles, which will have great impacts in designing beam scanning transmitarrays and conformal transmitarray designs.

Additionally, bandwidth improvement of transmitarray antennas is still an essential issue that needs further research. Very low profile transmitarray antennas using the receiver-transmitter design approach could achieve wideband performance with wideband elements, such as U-slot patches [50], and the use of tapered transmission lines [51] between the receiver and transmitter patches to control the transmission phase for wideband applications.

Moreover, for transmitarray antenna synthesis, inclusion of element transmission coefficient (magnitude and phase) in the optimization routine along with oblique incident considerations could potentially result in a better design tool leading to improved performance.

Furthermore, reconfigurable transmitarray antenna is an important direction that require further research and development. It has a great potential for applications where a scanning beam is desired.

APPENDIX A

S-matrix of Cascaded Layers

In order to obtain the S-parameters of the multilayer configurations, we should first develop the S-matrix of any two cascaded layers using the knowledge of the S-parameters of each individual layer as [1, 2],

$$S_{11}^C = \frac{S_{11}^2 S_{12}^1 S_{21}^1}{1 - S_{11}^2 S_{22}^1} + S_{11}^1, \tag{A.1}$$

$$S_{12}^C = S_{21}^C = \frac{S_{21}^1 S_{21}^2}{1 - S_{11}^2 S_{22}^1}, \tag{A.2}$$

$$S_{22}^C = \frac{S_{22}^1 S_{21}^2 S_{12}^2}{1 - S_{11}^2 S_{22}^1} + S_{22}^2, \tag{A.3}$$

where S_{11}^1, S_{12}^1, S_{21}^1, and S_{22}^1 are the S-parameters of the first layer, S_{11}^2, S_{12}^2, S_{21}^2, and S_{22}^2 are the S-parameters of the second layer, S_{11}^C, S_{12}^C, S_{21}^C, and S_{22}^C are the S-parameters of cascaded two layers.

Accordingly, the S-matrix of multiple-conductor layers separated by dielectric substrate can be computed (and hence the transmission coefficient S_{21}) by repeatedly cascading the S-parameters of the conductor layer defined in Equations (3.9) and (3.10) and the S-parameters of the dielectric substrate defined as [42],

$$S_{11} = S_{22} = \frac{\Gamma \left(1 - e^{-j2\beta L_d}\right)}{1 - \Gamma^2 e^{-j2\beta L_d}}, \tag{A.4}$$

$$S_{12} = S_{21} = \frac{\left(1 - \Gamma^2\right) e^{-j\beta L_d}}{1 - \Gamma^2 e^{-j2\beta L_d}}, \tag{A.5}$$

where

$$\Gamma = \frac{1 - \sqrt{\epsilon_r}}{1 + \sqrt{\epsilon_r}} \quad \text{and} \quad \beta = \frac{2\pi \sqrt{\epsilon_r}}{\lambda_0}.$$

It is worthwhile to notice that the S-matrix of the dielectric substrate is a function of the dielectric permittivity ε_r and the substrate thickness L_d, while the S-matrix of the conducting element layer is a function of its $\angle S_{21}$.

For the special case of air gap separation between layers rather than dielectric substrate, the S-parameters of Equations (A.4) and (A.5) are simplified by substituting $\varepsilon_r = 1$ to:

$$S_{11} = S_{22} = 0, \tag{A.6}$$

$$S_{12} = S_{21} = e^{-j\beta L_d}. \tag{A.7}$$

For the case of oblique incidence, both Γ and β have to be updated based on the analysis as discussed in [42].

Bibliography

[1] S. Datthanasombat, L. R. Amaro, J. A. Harrell, S. Spitz and J. Perret, Layered lens antenna, *IEEE Antennas and Propagation Society International Symposium*, pp. 777–780, Boston, 2001. DOI: 10.1109/aps.2001.959839. 3, 25, 33, 34, 57, 62, 69, 139

[2] A. H. Abdelrahman, F. Yang, and A. Z. Elsherbeni, Transmission phase limit of multilayer frequency selective surfaces for transmitarray designs, *IEEE Transactions on Antennas and Propagation*, vol. 62, no. 2, pp. 690–697, 2014. DOI: 10.1109/tap.2013.2289313. 3, 25, 33, 34, 36, 70, 96, 139

[3] R. Milne, Dipole array lens antenna, *IEEE Transactions on Antennas and Propagation*, vol. AP-30, no. 4, pp. 704–712, 1982. DOI: 10.1109/tap.1982.1142835. 33, 57, 69

[4] C. G. M. Ryan, M. Reza, J. Shaker, J. R. Bray, Y. M. M. Antar, and A. Ittipiboon, A wideband transmitarray using dual-resonant double square rings, *IEEE Transactions on Antennas and Propagation*, vol. 58, no. 5, pp. 1486–1493, 2010. DOI: 10.1109/tap.2010.2044356. 33, 39, 62, 69, 70, 95

[5] M. A. Al-Joumayly and N. Behdad, Wideband planar microwave lenses using subwavelength spatial phase shifters, *IEEE Transactions on Antennas and Propagation*, vol. 59, no. 12, pp. 4542–4552, 2011. DOI: 10.1109/tap.2011.2165515. 33

[6] M. Li, M. A. Al-Joumayly, and N. Behdad, Broadband true-time-delay microwave lenses based on miniaturized element frequency selective surfaces, *IEEE Transactions on Antennas and Propagation*, vol. 61, no. 3, pp. 1166–1179, 2013. DOI: 10.1109/tap.2012.2227444. 3, 33, 95

[7] D. M. Pozar, Flat lens antenna concept using aperture coupled microstrip patches, *Electronic Letters*, vol. 32, no. 23, pp. 2109–2111, 1996. DOI: 10.1049/el:19961451. 4, 95

[8] H. J. Song and M. E. Bialkowski, Transmit array of transistor amplifiers illuminated by a patch array in the reactive near-field region, *IEEE Transactions on Microwave Theory and Techniques*, vol. 49, no. 3, pp. 470–475, 2001. DOI: 10.1109/22.910550.

[9] P. Padilla de la Torre and M. Sierra-Castaner, Design and prototype of a 12-GHz transmitarray, *Microwave and Optical Technology Letters*, vol. 49, no. 12, pp. 3020–3026, 2007. DOI: 10.1002/mop.22950. 24

[10] H. Kaouach, L. Dussopt, R. Sauleau, and T. Koleck, X-band transmit-arrays with linear and circular polarization, *3rd European Conference on Antennas and Propagation, EuCAP*, pp. 1191–1195, Berlin, Germany, 2009. 69

[11] C.-C. Cheng, B. Lakshminarayanan, and A. A.-Tamijani, A Programmable lens-array antenna with monolithically integrated MEMS switches, *IEEE Transactions on Microwave Theory and Techniques*, vol. 57, no. 8, pp. 1874–1884, 2009. DOI: 10.1109/tmtt.2009.2025422. 24

[12] J. Y. Lau and S. V. Hum, A low-cost reconfigurable transmitarray element, *IEEE Antennas and Propagation Society International Symposium, APS-URSI*, South Carolina, s2009. DOI: 10.1109/aps.2009.5171561. 69

[13] H. Kaouach, L. Dussopt, R. Sauleau, and T. Koleck, Design and demonstration of an X-band transmit-array, *4th European Conference on Antennas and Propagation, EuCAP*, pp. 1–5, Barcelona, Spain, 2010. 24, 69

[14] R. Phillion, M. Okoniewski, Analysis of a transmit-array antenna for circular polarization, *26th Annual Review of Progress in Applied Computational Electromagnetics (ACES)*, pp. 804–807, Tampere, Finland, 2010.

[15] P. Padilla, A. M.-Acevedo, M. S.-Castañer, and M. S.-Pérez, Electronically reconfigurable transmitarray at Ku band for microwave applications, *IEEE Transactions on Antennas and Propagation*, vol. 58, no. 8, pp. 2571–2579, 2010. DOI: 10.1109/tap.2010.2050426. 24

[16] A. Clemente, L. Dussopt, R. Sauleau, P. Potier, and P. Pouliguen, Design of a reconfigurable transmit-array at X-band frequencies, *15th International Symposium on Antenna Technology and Applied Electromagnetics (ANTEM)*, Toulouse, France, 2012. DOI: 10.1109/antem.2012.6262295. 70, 95

[17] H. Kaouach, L. Dussopt, J. Lanteri, T. Koleck, and R. Sauleau, Wideband low-loss linear and circular polarization transmit-arrays in V-band, *IEEE Transactions on Antennas and Propagation*, vol. 59, no. 7, pp. 2513–2523, 2011. DOI: 10.1109/tap.2011.2152331. 69, 95

[18] J. Y. Lau and S. V. Hum, A wideband reconfigurable transmitarray element, *IEEE Transactions on Antennas and Propagation*, vol. 60, no. 3, pp. 1303–1311, 2012. DOI: 10.1109/tap.2011.2180475. 95

[19] A. Clemente, L. Dussopt, R. Sauleau, P. Potier, and P. Pouliguen, Wideband 400-element electronically reconfigurable transmitarray in X-band, *IEEE Transactions on Antennas and Propagation*, vol. 61, no. 10, pp. 5017–5027, 2013. DOI: 10.1109/tap.2013.2271493. 4, 24, 70, 95

[20] S. Kamada, N. Michishita, Y. Yamada, Metamaterial lens antenna using dielectric resonators for wide angle beam scanning, *IEEE Antennas and Propagation Society International Symposium APS-URSI*, Toronto, Canada, 2010. DOI: 10.1109/aps.2010.5561780. 5

[21] Q. Cheng, H. F. Ma, and T. J. Cui, Broadband planar Luneburg lens based on complementary metamaterials, *Applied Physics Letter*, vol. 95, no. 18, 2009. DOI: 10.1063/1.3257375. 5, 95

[22] Y. Zhang, R. Mittra, and W. Hong, On the synthesis of a flat lens using a wideband low-reflection gradient-index metamaterial, *Journal of Electromagnetic Waves and Applications*, vol. 25, no. 16, pp. 2178–2187, 2012. DOI: 10.1163/156939311798147015. 5

[23] M. Li and N. Behdad, Ultra-wideband, true-time-delay, metamaterial-based microwave lenses, *IEEE Antennas and Propagation Society International Symposium APS-URSI*, Chicago, IL, 2012. DOI: 10.1109/aps.2012.6349044. 5, 57, 95

[24] J. Huang and J. A. Encinar, *Reflectarray Antennas*, by Institute of Electrical and Electronics Engineers, John Wiley & Sons, 2008. DOI: 10.1002/9780470178775. 7, 30

[25] P. Nayeri, *Advanced Design Methodologies and Novel Applications of Reflectarray Antennas*, Doctoral Dissertation, Electrical Engineering Department, University of Mississippi, 2012. 7, 9, 11, 12, 13, 24

[26] T. B. A. Senior and J. L. Volakis, *Approximate Boundary Conditions in Electromagnetics*, by Institute of Electrical Engineers, Bookcraft Ltd, 1995. DOI: 10.1049/pbew041e. 9

[27] *CST Microwave Studio*, version 2012.01, 2012. xiii, 10, 34, 36, 43, 46, 53, 55, 57, 62, 70, 83, 90, 97, 107, 126

[28] P. Nayeri, A. Z. Elsherbeni, and F. Yang, Radiation analysis approaches for reflectarray antennas, *IEEE Antennas and Propagation Magazine*, vol. 55, no. 1, pp. 127–134, 2013. DOI: 10.1109/map.2013.6474499. 9, 11, 12, 59, 64, 99, 110, 116

[29] C. A. Balanis, *Antenna Theory Analysis and Design*, 3rd ed., Wiley, NY, 2005. 13, 63

[30] B. Devireddy, *Gain and Bandwidth Study of Reflectarray Antennas*, Master Thesis, Electrical Engineering Department, University of Mississippi, 2010. 13, 14, 15, 28, 29

[31] B. Devireddy, A. Yu, F. Yang, and A. Z. Elsherbeni, Gain and bandwidth limitations of reflectarrays, *Applied Computational Electromagnetics Society (ACES) Journal*, vol. 26, no. 2, pp. 170–178, 2011. 17, 20

[32] A. Yn, F. Yang, A. Z. Elsherbeni, J. Huang, and Y. R. Samii, Aperture efficiency analysis of reflectarray antennas, *Microwave and Optical Technology Letters*, vol. 52, no. 2, pp. 364–372, 2010. DOI: 10.1002/mop.24949. 17, 20, 21

[33] IEEE Standard # 145, Definitions of Terms for Antennas, *IEEE Transactions on Antennas and Propagation*, vol. AP-17, no. 3, pp. 262–269, 1969. DOI: 10.1109/tap.1969.1139442. 20

[34] D. M. Pozar, S. D. Targonski , and H. D. Syrigos, Design of millimeter—wave microstrip reflectarrays, *IEEE Transactions on Antennas and Propagation*, vol. 45, no. 2, pp. 287–296, 1997. DOI: 10.1109/8.560348. 28

[35] *Ansoft Designer*, version 6.1.0, Ansoft Corporation, Pittsburgh, PA, 2010. 33, 36, 39

[36] D. M. Pozar, *Microwave Engineering*, 3rd ed., John Wiley & Sons, Inc., New York, 2005. 34

[37] B. A. Munk, *Frequency Selective Surfaces, Theory and Design*, John Wiley & Sons, Inc., NY, 2000. DOI: 10.1002/0471723770. 35, 107

[38] M. Ando, T. Numata, J.-I. Takada, and N. Goto, A linearly polarized radial line slot antenna, *IEEE Transactions on Antennas and Propagation*, vol. 36, no. 12, pp. 1675–1680, 1988. DOI: 10.1109/8.14389. 61

[39] Y. Zhang, M. Abd-Elhady, W. Hong, and W. Li, Research progress on millimeter wave transmitarray in SKLMMW, *IEEE 4th International High Speed Intelligent Communication Forum (HSIC)*, Nanjing, China, 2012. DOI: 10.1109/hsic.2012.6212950. 95

[40] A. H. Abdelrahman, F. Yang, and A. Z. Elsherbeni, High gain and broadband transmitarray antenna using triple-layer spiral dipole elements, *IEEE Antennas and Wireless Propagation Letters*, vol. 13, pp. 1288–1291, 2014. DOI: 10.1109/lawp.2014.2334663. 95

[41] M. Li and N. Behdad, Wideband true-time-delay microwave lenses based on metallo-dielectric and all-dielectric lowpass frequency selective surfaces, *IEEE Transactions on Antennas and Propagation*, vol. 61, no. 8, pp. 4109–4119, 2013. DOI: 10.1109/tap.2013.2263784.

[42] C. A. Balanis, *Advanced Engineering Electromagnetics*, 2nd ed., Wiley, NY, 2012. 139, 140

[43] C. Guo, H. Sun, and X. Lu, Dual band frequency selective surface with double-four-legged loaded slots elements, *International Conference on Microwave and Millimeter Wave Technology (ICMMT)*, pp. 297–300, Nanjing, China, 2008. DOI: 10.1109/icmmt.2008.4540367. 95, 107

[44] L. Epp, C. Chan, and R. Mittra, The study of FSS surfaces with varying surface impedance and lumped elements, *IEEE Antennas and Propagation Society Symposium AP-S*, pp. 1056–1059, California, 1989. DOI: 10.1109/aps.1989.134882.

[45] S. M. Choudhury, M. A. Zaman, M. Gaffar, and M. A. Matin, A novel approach for changing bandwidth of FSS filter using gradual circumferential variation of loaded elements, *PIERS Proceedings*, pp. 1132–1134, Cambridge, 2010. 107

[46] P. Nayeri, F. Yang, and A. Z. Elsherbeni, Design of multifocal transmitarray antennas for beamforming applications, *IEEE Antennas and Propagation Society International Symposium, APS-URSI*, Florida, 2013. DOI: 10.1109/aps.2013.6711495. 115, 116

[47] P. Nayeri, F. Yang, and A. Z. Elsherbeni, Design and experiment of a single-feed quad-beam reflectarray antenna, *IEEE Transactions on Antennas and Propagation*, vol. 60, no. 2, pp. 1166–1171, 2012. DOI: 10.1109/tap.2011.2173126. 115, 116

[48] P. Nayeri, F. Yang, and A. Z. Elsherbeni, Design of single-feed reflectarray antennas with asymmetric multiple beams using the particle swarm optimization method, *IEEE Transactions on Antennas and Propagation*, vol. 61, no. 9, pp. 4598–4605, 2013. DOI: 10.1109/tap.2013.2268243. 115, 116, 117

[49] D. W. Boeringer and D. H. Werner, Particle swarm optimization vs. genetic algorithms for phased array synthesis, *IEEE Transactions on Antennas and Propagation*, vol. 52, no. 3, pp. 771–779, 2004. DOI: 10.1109/tap.2004.825102. 116

[50] K. F. Lee, K. M. Luk, K. F. Tong, S. M. Shum, T. Huynh, and R. Q. Lee, Experimental and simulation studies of the coaxially fed U-slot rectangular patch antenna, *IEE Proc. of the Microwave, Antennas and Propagation*, vol. 144, no. 5, pp. 354–358, 1997. DOI: 10.1049/ip-map:19971334. 137

[51] V. Demir, D. A. Elsherbeni, D. Kajfez, and A. Z. Elsherbeni, Efficient wideband power divider for planar antenna arrays, *Applied Computational Electromagnetics Society (ACES) Journal*, vol. 21, no. 3, pp. 318–324, 2006. 137

Authors' Biographies

AHMED H. ABDELRAHMAN

Ahmed H. Abdelrahman received B.S. and M.S. degrees from The Department of Electrical Engineering, Electronics and Communications, Ain Shams University, Cairo, Egypt, in 2001 and 2010, respectively, and he received a Ph.D. degree in engineering sciences from The Department of Electrical Engineering, University of Mississippi, University, MS, USA, in 2014. Dr. Abdelrahman is currently a research associate with the Antenna Research Group (ARG), in The Department of Electrical, Computer, and Energy Engineering, at The University of Colorado Boulder, Boulder, CO, USA. He also worked as a postdoctoral research associate for almost two years in The Department of Electrical and Computer Engineering at The University of Arizona, Tucson, AZ, USA. Additionally, he possesses over eight years of experience in Satellite Communications industry. He worked as a RF design engineer and a Communication System Engineer in building the low earth orbit satellite Egyptsat-1. His research interests include transmitarray/reflectarray antennas, mobile antennas, reconfigurable antennas, simultaneous transmit and receive (STAR) antennas, satellite communications, and thermoacoustic and millimeter-wave imaging.

Dr. Abdelrahman was the recipient of the several prestigious awards, including the Third-Place Winner Student Paper Competition Award at the 2013 ACES Annual Conference, and the Honorable Mention Student Paper Competition at the 2014 IEEE AP-S International Symposium on Antennas and Propagation.

FAN YANG

Fan Yang received B.S. and M.S. degrees from Tsinghua University, Beijing, China, in 1997 and 1999, respectively, and a Ph.D. degree from the University of California at Los Angeles (UCLA) in 2002.

From 1994–1999, he was a research assistant with the State Key Laboratory of Microwave and Digital Communications, Tsinghua University. From 1999–2002, he was a graduate student researcher with the Antenna Laboratory, UCLA. From 2002–2004, he was a post-doctoral research engineer and instructor with the Electrical Engineering Department, UCLA. In 2004, he joined the Electrical Engineering Department at The University of Mississippi as an assistant professor, and was promoted to a tenured associate professor. In 2011, he joined the Electronic Engineering Department at Tsinghua University as a professor, and has served as the Director of the Microwave and Antenna Institute since then.

Prof. Yang's research interests include antennas, surface electromagnetics, computational electromagnetics, and applied electromagnetic systems. He has published over 200 journal articles and conference papers, six book chapters, and three books entitled *Scattering Analysis of Periodic Structures Using Finite-Difference Time-Domain Method* (Morgan & Claypool, 2012), *Electromagnetic Band Gap Structures in Antenna Engineering* (Cambridge Univ. Press, 2009), and *Electromagnetics and Antenna Optimization Using Taguchi's Method* (Morgan & Claypool, 2007).

Prof. Yang served as an associate editor of the *IEEE Transactions on Antennas and Propagation* (2010–2013) and an associate editor-in-chief of *Applied Computational Electromagnetics Society (ACES) Journal* (2008–2014). He was the Technical Program Committee (TPC) Chair of *2014 IEEE International Symposium on Antennas and Propagation and USNC-URSI Radio Science Meeting*. Dr. Yang has been the recipient of several prestigious awards and recognitions, including the Young Scientist Award of the 2005 URSI General Assembly and of the 2007 International Symposium on Electromagnetic Theory, the 2008 Junior Faculty Research Award of the University of Mississippi, the 2009 inaugural IEEE Donald G. Dudley Jr. Undergraduate Teaching Award, and the 2011 Recipient of Global Experts Program of China.

ATEF Z. ELSHERBENI

Atef Z. Elsherbeni received an honor B.Sc. degree in electronics and communications, an honor B.Sc. degree in applied physics, and a M. Eng. degree in electrical engineering, all from Cairo University, Cairo, Egypt, in 1976, 1979, and 1982, respectively, and a Ph.D. degree in electrical engineering from Manitoba University, Winnipeg, Manitoba, Canada, in 1987. He started his engineering career as a part time Software and System Design Engineer from March 1980 to December 1982 at the Automated Data System Center, Cairo, Egypt. From January to August 1987, he was a post doctoral fellow at Manitoba University. Dr. Elsherbeni joined the faculty at the University of Mississippi in August 1987 as an assistant professor of electrical engineering. He advanced to the rank of associate professor in July 1991, and to the rank of professor in July 1997. He was appointed as Associate Dean of Engineering for Research and Graduate Programs from July 2009 to July 2013 at the University of Mississippi. Dr. Elsherbeni joined the Electrical Engineering and Computer Science Department at Colorado School of Mines in August 2013 as the Dobelman Distinguished Chair Professor. He currently is the Electrical Engineering Division Director. He spent a sabbatical term in 1996 at the Electrical Engineering Department, University of California at Los Angeles (UCLA) and was a visiting Professor at Magdeburg University during the summer of 2005 and at Tampere University of Technology in Finland during the summer of 2007. In 2009, he was selected as Finland Distinguished Professor by the Academy of Finland and TEKES.

Over the years, Dr. Elsherbeni participated in acquiring millions of dollars to support his research group activities dealing with scattering and diffraction of EM waves by dielectric and metal objects, finite difference time domain analysis of antennas and microwave devices, field visualization and software development for EM education, interactions of electromagnetic waves with human body, RFID and sensor Integrated FRID systems, reflector and printed antennas and antenna arrays for radars, UAV, and personal communication systems, antennas for wideband applications, and measurements of antenna characteristics and material properties. He is the co-author of the books *Antenna Analysis and Design Using FEKO Electromagnetic Simulation Software*, ACES Series on Computational Electromagnetics and Engineering, SciTech 2014, *Double-Grid Finite-Difference Frequency-Domain (DG-FDFD) Method for Scattering from Chiral Objects*, Morgan & Claypool, 2013, *Scattering Analysis of Periodic Structures Using Finite-Difference Time-Domain Method*, Morgan & Claypool, 2012, *Multiresolution Frequency Domain Technique for Electromagnetics*, Morgan & Claypool, 2012, *The Finite Difference Time Domain Method for Electromagnetics with Matlab Simulations*, Scitech, 2009 1st ed., and 2016 2nd ed. by IET, *Antenna Design and Visualization Using Matlab*, Scitech, 2006, *MATLAB Simulations for Radar Systems Design*, CRC Press, 2003, *Electromagnetic Scattering Using the Iterative Mul-*

tiregion Technique, Morgan & Claypool, 2007, *Electromagnetics and Antenna Optimization using Taguchi's Method*, Morgan & Claypool, 2007, *Scattering Analysis of Periodic Structures Using Finite-Difference Time-Domain Method*, Morgan & Claypool, 2012, *Multiresolution Frequency DomainTechnique for Electromagnetics*, Morgan & Claypool, 2012, and the main author of the chapters "Handheld Antennas" and "The Finite Difference Time Domain Technique for Microstrip Antennas" in *Handbook of Antennas in Wireless Communications*, CRC Press, 2001.

Dr. Elsherbeni is a Fellow member of the Institute of Electrical and Electronics Engineers (IEEE) and a fellow member of The Applied Computational Electromagnetic Society (ACES). He is the Editor-in-Chief of the *ACES Journal*, and a past Associate Editor of the *Radio Science Journal*. He was the Chair of the Engineering and Physics Division of the Mississippi Academy of Science and was the Chair of the Educational Activity Committee for the IEEE Region 3 Section. He held the president position of ACES Society from 2013–2015.

PAYAM NAYERI

Payam Nayeri received a B.Sc. in applied physics from Shahid Beheshti University, Tehran, Iran, in 2004, an M.Sc. in electrical engineering from Iran University of Science and Technology, Tehran, Iran, in 2007, and a Ph.D. in electrical engineering from The University of Mississippi, University, MS, USA, in 2012.

From 2008–2013, Dr. Nayeri was with the Center for Applied Electromagnetic Systems Research (CAESR) at The University of Mississippi. Prior to this, he was a visiting researcher at the University of Queensland, Brisbane, Australia. From August 2012 to December 2013 he was a postdoctoral research associate and instructor with the Electrical Engineering Department, The University of Mississippi. From January 2014 to June 2015, he was a post-doctoral fellow with the Electrical Engineering and Computer Science Department, Colorado School of Mines, Golden, CO, USA.

Dr. Nayeri joined the Electrical Engineering and Computer Science Department at Colorado School of Mines as an assistant professor in July 2015. His current research is in the areas of multifunctional and adaptive microwave circuits, wireless energy harvesting, active and wideband antenna arrays, and adaptive beamforming. Dr. Nayeri is a member of *IEEE*, Sigma Xi, and Phi Kappa Phi, and has authored over 60 journal articles and conference papers. He has been the recipient of several prestigious awards, including the *IEEE Antennas and Propagation Society* Doctoral Research Award in 2010, the University of Mississippi Graduate Achievement Award in Electrical Engineering in 2011, and the Best Student Paper Award of the 29th International Review of Progress in ACES.

Printed in the United States
by Baker & Taylor Publisher Services